SUSTAINABLE
COMMUNITY
DEVELOPMENT

Studies in Economic, Environmental, and Cultural Revitalization

SUSTAINABLE COMMUNITY DEVELOPMENT

Studies in Economic, Environmental, and Cultural Revitalization

Edited by
Marie D. Hoff

Lewis Publishers

Boca Raton   Boston   London   New York   Washington, D.C.

Library of Congress Cataloging-in-Publication Data

Catalog information may be obtained from the Library of Congress.

Table of Contents

Sustainable Regional Community Development Cases

Acknowledgments

Grateful acknowledgment is given to The Foundation for Community Encouragement (P.O. Box 17210, Seattle, WA 98107) for permission to quote "Some Rules of Community Building" and "Some Principles of Community" (in Chapter 9).

Grateful acknowledgment is also given to The Oregon Business Council (1100 SW 6th Avenue, Suite 1608, Portland, OR 97204-1090) for permission to reproduce "Characteristics of the Industrial and Natural Economies" (in Chapter 1).

I thank Dennis McClellan, of CRC Press, for recognizing the importance of this set of studies, and Sandy Pearlman, also at CRC Press, for assistance in the process of preparing this manuscript. Rosemary Crabtree and Laurie Byrd, "Secretaries Extraordinaire" at Boise State University, gave invaluable assistance on the intricacies of word-processing and e-mail translation. John Fritz, BSU student, gave generously of his time to assist with proofreading.

I thank numerous friends and colleagues around the United States, and several from other countries, who helped me locate communities engaged in sustainable development efforts. I am also grateful to my colleagues Dr. John G. McNutt of Boston College, who read and offered helpful comments on Chapter 1 of this volume, and Dr. Mary E. Rogge of the University of Tennessee, who gave valuable feedback on the concluding chapter. I am very appreciative of all the authors who participated so enthusiastically in this project to disseminate stories of communities of hope and resilience.

Most of all, I thank the thousands of people whose stories are told in this volume, for their diligence and persistence in looking for ways to cherish their local environment and to love and foster the development of people who live and work in those special places. This book is dedicated to them.

Marie D. Hoff

Editor

 **Marie D. Hoff** is Associate Professor of Social Policy and Community Organization Practice, School of Social Work, Boise State University, Boise, Idaho. She also has served on the faculty at Saint Louis University. Dr. Hoff holds a Ph.D. degree in social welfare from the University of Washington (1987). She has conducted research and published evaluations of social programs, community organization practice models, and social action by rural women. Her previously published books include (with co-editor John G. McNutt) *The Global Environmental Crisis: Implications for Social Welfare and Social Work* (1994) and (with co-editor Gerard Magill) *Values and Public Life* (1995). She is active in community-based initiatives regarding environmental protection and social justice, and she has lectured on the same in interdisciplinary and community-based educational settings.

Contributing Authors

Mark Bekkering, M.A., is Senior Policy Analyst, Regional Environmental Department, Regional Municipality of Hamilton–Wentworth, Ontario, Canada. He is an environmental planner and for the past four years has had responsibility for coordination and implementation of the primary projects making up the Hamilton–Wentworth Sustainable Community Initiative.

David R. Cox, Ph.D., is Director of the Regional Social Development Centre and Professor of Social Work, Faculty of Social Sciences, La Trobe University, Bundoora, Victoria, Australia. Dr. Cox's work focuses on poverty, migration, and development in the Asia–Pacific region.

Sandra De Carlo, M.Sc., is a researcher at Instituto Brasileiro de Geografia e Estatística, Rio de Janeiro, Brazil. She is an economist with specialization in environmental studies.

José Drummond, M.Sc., is Adjunct Professor of Political Science at Universidade Federal Fluminense, Rio de Janeiro, Brazil. He is also a Ph.D. candidate in land resources studies at the University of Wisconsin, Madison.

John Eyles, Ph.D., is Chair of the Environmental Health Program and Director of the McMaster Institute of Environment and Health, McMaster University, Hamilton, Ontario, Canada. He is also Professor of Geography at McMaster and has associate status with the Departments of Sociology and Clinical Epidemiology and Biostatistics and the Centre for Health Economics and Policy Analysis.

Frank Fromherz, Ph.D., is the Director of the Office of Justice and Peace and the Campaign for Human Development, Archdiocese of Portland in Oregon. He is also Adjunct Professor of Religion and Society, Theology Department, University of Portland.

Kirk Johnson, M.P.A., was, while conducting this study, a Policy Analyst with the Northwest Policy Center, Graduate School of Public Affairs, University of Washington, Seattle.

Joan Legg, M.A., is the Director of the Southern Oregon Economic Development Coalition, Medford, Oregon. She has 35 years of experience in community organizing and direct service with the poor and serves on the National Advisory Committee of the Campaign for Human Development.

Michelle Livermore, M.S.W., is Research Associate in the Office of Research and Graduate Studies at Louisiana State University, Baton Rouge, Louisiana. She is also a doctoral candidate at Louisiana State University in the field of sociology.

Mark W. Lusk, Ed.D., is Director of International Programs at the University of Montana, Missoula. He has worked in Kazakstan for various projects funded by the U.S. Agency for International Development, the World Bank, and the Japan International Cooperation Agency. He specializes in agriculture education and rural development.

Jon Kei Matsuoka, Ph.D., is Professor of Social Work, University of Hawaii, Manoa. He has conducted numerous studies on rural Hawaii and other Pacific Island communities. His areas of research interest and experience include culture, mental health, social change, human service delivery systems, and economic development.

Davianna Pomaika`i McGregor, Ph.D., is Associate Professor of Ethnic Studies, University of Hawaii, Manoa. Trained as a historian of Hawaii and the Pacific, her research has focused on the continuity of Hawaiian cultural, spiritual, and subsistence beliefs, customs, and practices in rural Hawaiian communities.

James Midgley, Ph.D., is Harry and Riva Specht Professor of Public Social Service and Dean of the School of Social Welfare at the University of California-Berkeley. He was formerly the Associate Vice-Chancellor of the Office of Research and Economic Development at Louisiana State University, Baton Rouge. He has published extensively in studies of social welfare and social development.

Luciano Minerbi, Dr.Arch., MUP, AICP, AAIA, is Professor of Urban and Regional Planning, University of Hawaii, Manoa. His consulting and research experience for international, national, and local agencies and organizations has been in the field of environmental information, planning, and management.

S.I. Ospanov is Professor of Sociology at the Kazak State University in Almaty, Kazakstan. His areas of specialization are rural sociology and sociological theory. A graduate of Moscow State University, he has conducted extensive field research on agricultural and rural development in Kazakstan. He has consulted for the World Bank, Harza Engineering, and USAID on farmer organization and agricultural development.

Mary E. Rogge, Ph.D., is Assistant Professor of Social Work in the College of Social Work, University of Tennessee, Knoxville. Her research interests are in the areas of justice in environmental and social welfare issues, citizen participation and community organization, children and toxic chemical risk, and the effects of technological and natural hazards on human populations.

Jonathan M. Scherch, Ph.D., is Assistant Professor of Social Work at Pacific Lutheran University, Tacoma, Washington. At the time of conducting this research, he was a doctoral candidate at the University of Tennessee, Knoxville, and adjunct faculty member at Lincoln Memorial University in Harrogate, Tennessee. He is a returned United States Peace Corps Volunteer (Jamaica, 1991–93).

Introduction

Marie D. Hoff

This book is a collection of case studies of a variety of urban and rural communities which have committed themselves to a process of sustainable development. The collection includes international examples as well as urban and rural cases from the United States. Each chapter includes two major sections: a narrative of the processes and strategies which led to revitalization and an analysis of the strengths, limitations, and replicability of their methods and approaches. The distinctive focus of this collection is the emphasis on communities which have adopted — or at least are working toward — harmonizing environmental sustainability with their economic and social development efforts and goals.

The 1990s has been marked by a widespread awareness of the convergence of environmental, economic, and social problems and issues. Many local workers have begun to recognize that severe setbacks or even collapse of their local economy is strongly related to environmental problems — either to the depletion of local resources (such as timber, fish, or minerals) or to severe pollution and degradation of the local ecosystem. Local economic collapse frequently is related strongly to changing circumstances in the turbulent global economy. Many communities have rallied creatively in response to the departure of major corporate employers that have either gone out of business or have relocated their corporate resources elsewhere. Luring a large corporation to offer new jobs is simply not a realistic option for many communities. In some cases, because of consciousness-raising education and a change of values, members of the community agree that this traditional strategy is no longer a desirable route to economic well-being.

Out of the roots of despair and decline, disparate groups in many local communities have come together to devise and develop new ways to live responsibly in their local ecosystem, amid challenging new socioeconomic conditions. Many of these grass-roots and even region-wide projects are characterized by a democratic participatory approach and also include strong attention to restora-

1-57444-129-9/98/$0.00+$.50
© 1998 by CRC Press LLC

tion and revitalization of cultural institutions and practices. In the jargon of development theory, this attention to participation and culture is often referred to as "the human factor" in development. In this collection, the human, cultural aspect of development includes examples such as the following: evidence of improved social relationships and decreased conflict among groups in the local community, new or restored community customs and rituals of celebration and leisure activity, new socialization and skills-building opportunities for youth, and new or expanded leadership and skills development for adults who were deprived of these opportunities in a former economic regime. Culture also includes learning new or different ways to utilize environmental resources for human needs in patterns and levels that are environmentally sustainable. The case studies also address new or revitalized cultural practices with respect to how work is organized and carried out.

In-depth analysis of successful community-based economic revitalization and development is sparse. Examples of communities which have explicitly attempted to integrate the economic, the environmental, and the human factor are even more rare. It is difficult to locate case studies of community development which are detailed enough to assist the reader in understanding the actual activities which were undertaken by a cadre of people who instigated a development process. (In contrast, there exists an extensive body of literature on the theoretical bases of community-based economic development.) Many local community participants in these efforts also lack convenient access to stories about the similar conditions and efforts of citizens in other communities, especially international similarities. It is true that case examples of community development successes are often cited in longer, book-length monographs on one community's experience (such as the several book-length studies of the Mondragon cooperatives in the Basque region of Spain). Examples of community development are also cited in newsletters of activist organizations. However, this type of citation is usually too brief to really assist the reader in understanding the processes that aided or retarded the success of the case cited. This collection of chapter-length case examples is intended to provide this more detailed description of the process and to demonstrate the commonalities in approach across a wide variety of environmental and cultural settings.

This book is targeted both to students and to community development activists and experts who are engaged in economic and environmental restoration in a local setting. The three-pronged approaches reported (i.e., the environmental, economic, and cultural) are relevant to a wide variety of academic fields of study — most notably interdisciplinary environmental studies programs, special courses in alternative community-based economic development methodology, programs and courses in international development, community organization courses, and community planning courses in social work. Potentially, ethnic studies programs will also find the book useful, as a number of the communities include oppressed ethnic groups, such as North and South American Indian groups, and countries devastated by modern industrial development, such as Kazakstan and the Philippines. Community development organizations, foundation or institute staff, and staff of local, state or province, and national government economic development and planning agencies may also find the studies useful, as most of the chapters address the role of various agencies and levels of government in the development process.

Chapter 1 introduces the case studies by reviewing the theoretical and historical foundations of sustainable community development and its currently developing strategic alliance with environmental activism. The concluding chapter delineates some of the key common and distinctive themes, to assist the reader to see how the communities did follow many similar principles and fundamental approaches to social change, even though they were applied in distinctive patterns and with varying levels of success. Each community's story told here is a work in progress, which also provides a realistic picture of the difficulties which can be expected. Most importantly, this edited volume is intended to engender hope in the creative capacities of people to promote social development and economic security for all people within a framework of respect for the natural limits of our global environmental heritage.

CHAPTER 1

Sustainable Community Development: Origins and Essential Elements of a New Approach

Marie D. Hoff

In the past decade, the concept of sustainable development has swept through many social movements around the world. Strategic action for sustainability is occurring in the fields of economic development as well as governmental policy and program planning for urban and rural social development. The environmental movement in countries of both the North and South and the international women's development movement also have adopted sustainability as a guiding principle for action. Yet, definitions of sustainability and effective approaches to achieve it are unclear. Sustainable development "is in real danger of becoming a cliché like appropriate technology — a fashionable phrase that everyone pays homage to but nobody cares to define" (Lélé, 1991, p. 607). The expanding collection of writing about sustainable development reveals that the concept raises important questions about reigning cultural paradigms and values, including the assumptions and activities of modern economics, science, and technology.

Some of the history and key components of debate in the conversation about sustainability are discussed in this chapter. Environmental advocates stress environmental sustainability, while workers and economic development experts focus on economic sustainability, and those in human development work stress cultural and social sustainability. The purpose of this book, and of this introductory discussion, is to demonstrate what many development practitioners already recognize as the leading challenge to action in the decades ahead, namely, the integration and harmonization of cultural, economic, political, and environmental factors in the process of sustainable development.

DEFINITIONS AND HISTORY OF SUSTAINABLE DEVELOPMENT

The Scottish Natural Heritage Society (Hardy and Lloyd, 1994) articulated a number of principles that should inform sustainable development:

> These are: non-renewable resources should be used wisely and sparingly at a rate which does not restrict the options of future generations (wise use); renewable resources should be used within the limits of their capacity for regeneration (carrying capacity); the quality of the natural heritage as a whole should be maintained and improved (environmental quality); in situations of great complexity or uncertainty society should act in a precautionary manner (precautionary approach); and there should be an equitable distribution of the costs and benefits (material and non-material) of any development (shared benefits) (p. 776).

Williams and Haughton (1994) reduce these principles for action to three: intergenerational equity, social justice (includes intragenerational equity in distribution of resources and participation in planning), and transfrontier responsibility (global environmental stewardship) (p. 116). Sustainable development is potentially measurable, in the attributes listed by economists Pearce et al. (1990): non-decrease of natural stock over time (p. 1) and increases or improvements in measures of human well-being, such as income, education, health, and basic freedoms (p. 2). Daly (1996) stresses the importance of distinguishing growth (quantitative increase) from development (qualitative improvements).

The most well-known definition of sustainable development is that of the World Commission on Environment and Development (WCED) (1987): "...development that meets the needs of the present without compromising the ability of future generations to meet their own needs" (p. 43). Implied in this simple definition are two controversial concepts: the idea of needs implies priorities and judgment on the equitable use of resources, and it implies that limitations exist — earth's natural resources and our right to use them are not infinite. Controversy over use and preservation of natural resources has been played out in contemporary times in the tension between the environmental protection movement and economic development.

Sustainable Development and the Environmental Movement

The modern environmental movement is frequently traced to Rachel Carson's publication *Silent Spring* (1962), which demonstrated the harmful effects of synthetic chemicals on biological systems. In the wealthier countries of the North, especially the United States, the mainstream environmental movement is identified in the public eye with the conservation movement to set aside protected natural areas (which actually had begun already in the nineteenth century) and with the more recent efforts to reduce air and water pollution through governmental policy and regulation of industry. "Public opinion polls in recent years [in the United States] have consistently indicated a broad popular shift toward environmental values" (Shabecoff, 1993, p. 247).

Another landmark study, *The Limits to Growth* (Meadows et al., 1974), linked global environmental concerns to poverty, malnutrition, and the world political structures, and it challenged the prevailing assumption in economic development circles that unlimited economic growth is possible in a world of finite natural resources. Publication of *Our Common Future* (WCED, 1987), a study commissioned by the United Nations, represented another important step in establishing the linkages between environmental depletion and prospects for economic and social development. This study articulated clearly the moral imperative to pass on an intact, viable environment to future generations. One might reasonably argue that this imperative is the very nucleus of the idea of sustainable development.

The international environmental movement has forced developers, economists, governments, and ordinary people to examine in-depth the science, values, and goals of social and economic development. The mainstreaming of sustainable development is evident from its incorporation into the agenda of the United Nations, the adoption of its terminology into the goals of international organizations such as the World Bank, and the establishment of numerous governmental organizations, such as the President's Council on Sustainable Development (1996) in the United States.

However, the so-called mainstream environmental movement cannot take primary credit for the growing recognition that efforts to protect the environment must incorporate the economic and cultural survival of people. Credit for this new awareness of the human factor must go primarily to poor, minority, and indigenous peoples — in both the North and South — who have led popular movements to protest a number of environmental injustices. During the past two decades, such groups have mounted protests against the use of poor communities as dumping grounds for toxic wastes or as sites for location of dangerous industries (Bullard, 1990, 1993; Commission for Racial Justice, 1987; Hofrichter, 1993; Rogge, 1994). Poor communities — especially in the South — have become more articulate and have gained greater international attention in their struggles against extreme economic exploitation, use of their lands as dumping grounds for toxic wastes, and depletion of their natural resources for the international market. The movement for environmental justice among indigenous peoples, and among poor and working-class citizens, has raised significant challenges to the mainstream environmental movement, which has sometimes been accused of being elitist, indifferent, and possibly even hostile to consideration of human needs in relationship to environmental protection. In recent years, mainstream environmental organizations have attempted to become more responsive to the concerns of poor and minority groups, although with limited improvement (Shabecoff, 1993, pp. 263–264).

However, the growing strength of poor and minority groups in the international environmental protection debate can be inferred from the significant

voice they found at the 1992 world summit on the environment in Rio de
Janeiro. Here, representatives of the Northern mainstream environmental protection movement, governmental and economic development interests, and poor
people's movements met, clashed, dialogued, and reached some uncertain level
of understanding that neither economic development nor environmental protection can be achieved without consideration for the welfare of people at the
grass-roots levels of society (Shabecoff, 1996, pp. 60–77).[1] Poor people's
movements in both the wealthier and poorer countries of the globe have played
a significant role in advancing the notion of sustainability to include the harmonization and integration of environmental protection with economic and social
well-being.

Certain segments of the environmental movement have focused on world
population growth as the key to environmental decline. The international women's
movement has insisted that social and economic equality for women is the key
to population control. The 1994 global summit on population, held in Cairo, was
remarkable, after many years of limited attention to women's development
concerns, in finally garnering global-level acknowledgment of the centrality of
women's economic and social development as the key ingredient in population
control. The concerns of the poor, and especially of women, have had significant
influence in generating greater awareness of the linkages between environmental issues and the global economic system. Women have been key players in
demonstrating that economic development cannot succeed without attention to
human development. These human factors include access to healthcare, education, political voice and office, protected legal rights, and property rights.

Sustainability and Modern Economic Development

In classical philosophy, economics was a branch of ethics and the central question was the justice of economic distribution and exchange. Since Adam Smith,
the central questions of economics have been (1) under what conditions the
maximum increase in wealth is generated and (2) what the role of government
is in this process. Smith's laissez-faire philosophy more or less prevailed until
the Great Depression and World War II. After these transforming events, Keynes'
prescriptions for governmental guidance and management of macro-economic
factors (finance, trade, taxation, maintenance of competitive markets, etc.) were
widely accepted and guided the advanced industrialized countries to unprec-
edented levels of wealth and widespread prosperity (Barber, 1967; Keynes,
1964). It appeared that the extreme boom-and-bust cycles of capitalism had
indeed been tamed.

However, now the global flow of capital to countries with low wages and
low levels of government regulation has raised anew the issue of the proper and
effective role of government in directing the economy and ensuring the social
well-being of the citizenry. (Also, until fairly recently, the responsibility of
government to protect and regulate use of the natural environment has been

largely developed in a separate policy track from its role with respect to economic management.) According to Midgley (1995), the government's role in promoting social welfare (in many of the industrialized, capitalistic countries, especially the United States) has been developed separately from economic policy.[2]

The increasing maldistribution of wealth in the United States has led some critics to reexamine this lack of integration between social and economic policy. Nevertheless, the prevailing opinion in the United States appears to support the conservative argument that capitalist markets, free from government constraints, are the best mechanism to promote economic growth and wealth accumulation. Social problems are still largely viewed as a separate issue. The failed Communist experiment as an archetype of government control of economic activity, along with increased exposure of the dreadful political terror imposed by authoritarian governments of all stripes, have strengthened American belief in free markets and minimalist government. "The dilemma [of state control vs. free enterprise] arises in how to preserve a liberal economic order while at the same time preserving the natural resource base for future generations, and requiring private entities to take responsibility for the external diseconomies that they generate" (Mikesell, 1992, p. 76).

Midgley (1995) notes that in the poorer countries of the world, many of which were newly emancipated from the shackles of colonial control and exploitation after World War II, different concerns and emphases prevailed with respect to the role of government and the relationship between social and economic development. In his analysis, many such countries endeavored to integrate economic development with improvements in the social welfare of the population (i.e., with measurable progress in literacy, educational achievement, health and mortality indicators, sanitary conditions, etc.). The annual *Human Development Report* from the United Nations confirms that genuine improvements in the lives of millions of people have been realized through this strategy. However, the same annual report, and numerous other sources, confirm that approximately one billion people (about 20% of the world population) exist in absolute hunger and destitution within a depleted, polluted natural environment. Causes of this deplorable situation include cultural factors, as well as incorporation of Third World countries into the global political and economic system and lack of sustainable environmental policy and practice (Athanasiou, 1996).

In recent years, the mounting international debt of poor nations has come to symbolize the injustice and failure of mainstream economic development and has become a central variable in the struggle over Third World environmental resources.[3] Economists associated with international financial institutions, such as the World Bank and the International Monetary Fund, have frequently advocated a decoupling of economic and social strategies in the "structural adjustment programs" imposed on hard-pressed, indebted Third World governments in exchange for loans (Korten, 1995). More recently, structural adjustment programs have also been imposed on Eastern European countries, newly emancipated from the former Soviet Union. In the past decade, intensive advocacy research has drawn attention to the ecological and social devastation wrought by structural adjustment policies (Athanasiou, 1996, pp. 148–156; Brecher and Costello, 1994; Rich, 1994).

Mainstream economic development policy for developing nations has tended to emphasize large-scale economic development projects, modeled on the Western ideology and path of modern progress and growth (Daly, 1996). These include infrastructure projects, such as construction of hydroelectric dams or roads, extraction of natural resources such as wood and minerals, and agricultural production for trade on the international market. Local resistance and global political pressure by social and environmental action groups have caused these global institutions to reexamine their traditional approaches.[4]

Community-Based Economic Development Movement

In the United States, the movement for community-based economic development began during the War on Poverty in the 1960s and expanded among working-class communities in the late 1970s, when many manufacturing workers began to realize that large corporate employers were not going to be coming back to provide jobs. Slowly, groups emerged in many communities to assess their remaining social and environmental resources and to implement new, locally rooted approaches to economic organization and development[5] (Betancur et al., 1991; Bruyn and Meehan, 1987; Gunn and Gunn, 1991; Rubin and Rubin, 1992, pp. 414–436). Increasingly scarce resources and wider knowledge about environmental threats of all kinds have gradually led leaders and activists in many communities to incorporate environmental sustainability into their social and economic redevelopment efforts (President's Council, 1996).

Community-based economic development in the South has grown from many complex historical and social forces, such as anti-colonialism and national independence movements after World War II, the leadership of the United Nations in fostering integration between social and economic development (Midgley, 1995), the global feminist movement and its advocacy for women in development (Development Alternatives with Women for a New Era, 1995; Sen and Grown, 1987), and, since the 1980s, a plethora of local efforts to find alternative economic models to counter the pauperization engendered by the global capitalist market. Worker cooperatives and other forms of community ownership and control of enterprises constitute the core feature of community economic development. As noted earlier, another central aspect of community-based economic development is production for local needs first, with surplus for export being subsidiary to this goal.

In summary, economic decline, local communities' struggles to recover and develop new models of economic productivity, and the environmental concerns of disadvantaged groups have catalyzed current efforts to invent models of sustainable development that address economic and environmental concerns explicitly. Sustainable development has been adopted as a goal by many community-based and regional economic development groups in various countries.[6]

Nevertheless, the concept of sustainable development — and clear knowledge of how to achieve it — remains elusive. The theoretical and practical issues include sociocultural concerns, economic productivity and the justice of economic distribution, and policy, planning, and governance questions.[7]

SOCIAL AND CULTURAL VALUES AND BEHAVIORS

The assertion that achievement of sustainable patterns of economic development will require deep changes in many dominant social values is quite threatening to many people in advanced industrialized societies. The primary dedication to consumerism and private accumulation of material goods must change to a new emphasis on the values of adequate, but modest, satisfaction of the basic needs of all and the cultivation of non-material goods, such as leisure and community interaction, family and friendship, development of arts and personal skills, and so forth (Durning, 1992; Hoff, 1994; Wachtel, 1989). Put another way, communities working on sustainability initiatives must address how their economic enterprises contribute to meeting basic needs (i.e., housing, food, energy, health, education, and transit) and how they are balanced with policies to promote local quality of life while sustaining the ecological base. For example, a community might choose to develop affordable mass transit which would also reduce environmental threats and, potentially, increase time and financial resources for personal development and community-building activities.[8]

In normative terms, sustainable development requires a deemphasis on competition and the libertarian values of unfettered individual choice, which owe no accountability for effects on the natural environment or the choices of other people. Achievement of sustainability will require values of cooperation and democratically developed community consensus for action (Sale, 1985). Fundamental human and civil rights must always be respected, but, for example, personal use of property may be subject to a greater degree of community sanction. In the United States, perhaps more than in any other country, the sacrosanctness of private property rights is the value which most impedes community action for environmental protection. Taxation for social spending is viewed by some as theft of private property, which also thwarts societal response to common human needs.

Participatory planning is essential to development of the values which must undergird sustainable development: without substantial engagement in the process of *conscientization* (consciousness raising, critical thinking) and decision making, local community members are unlikely to change their values from materialism, which lionizes consumption and private property rights, to a concern for quality of life, social justice, and the common good (Daly and Cobb, 1989). Participation in decision making also helps foster the trust in one another which is necessary for a willingness to subscribe to communal (vs. individualistic) solutions to problems and needs. Moreover, without participation in the process of setting priorities and choosing goals, people in societies around the globe are increasingly unwilling to accept strategies imposed by elitist, bureaucratic, or authoritarian powers.

While sustainable development requires sociocultural changes in values and behavior, it also requires positive valuation of cultural continuity and social stability for the many and diverse human societies. Myriad modern forces have contributed to the sense of urgency behind the goals of fostering cultural continuity. Members of the environmental movement have realized that ecological breakdown and attendant economic collapse destroy human communities as well as the natural ecosystem. Researchers and workers in international development have studied cultural breakdown, which results from impoverishment, war, and violence, and the massive increases in refugees in many parts of the globe. In complex and varied patterns, the destabilization and sheer destruction of cultures are traceable to unsustainable environmental and economic policies (Athanasiou, 1996; George, 1992; Korten, 1995; Rich, 1994). An analogy can be seen between the weakening of an ecosystem through reductions in biodiversity and the weakening of human communities through destruction of cultures. Just as each individual species supports the stability and continuity of the ecosystem, one can argue that diverse cultures support the survival and the enrichment of human society.

Science and Technology

Logical positivism, the philosophical foundation of modern empirical science, posits that science is a value-free or value-neutral enterprise. The sometimes negative environmental effects of technology (i.e., as applied science) have raised challenges to the proposition of scientific value-neutrality. Ecological science, which studies not the isolated, separated components, but the interrelated, interdependent structure and functioning of living, dynamic systems, demonstrates the value implications of various approaches to science and technology. The book *Small Is Beautiful: Economics As If People Mattered,* by English economist E.F. Schumacher (1973), was a pathbreaking study which initiated an ongoing debate in the field of development over the proper uses and scale of technology. Numerous research studies of the unforeseen negative environmental effects of large-scale technological projects, such as dams and irrigation projects, have heightened awareness among many people of the value-ambiguity and scientific validity of technological innovation. "Progress," that is, the unexamined boosterism of twentieth century science and industry, is no longer viewed as an unalloyed good.

The fact that technological progress sometimes results in greater social inequities and disruption of cultural customs has also contributed to awareness of its value dimensions. For example, Indian economist Vandana Shiva (1991)

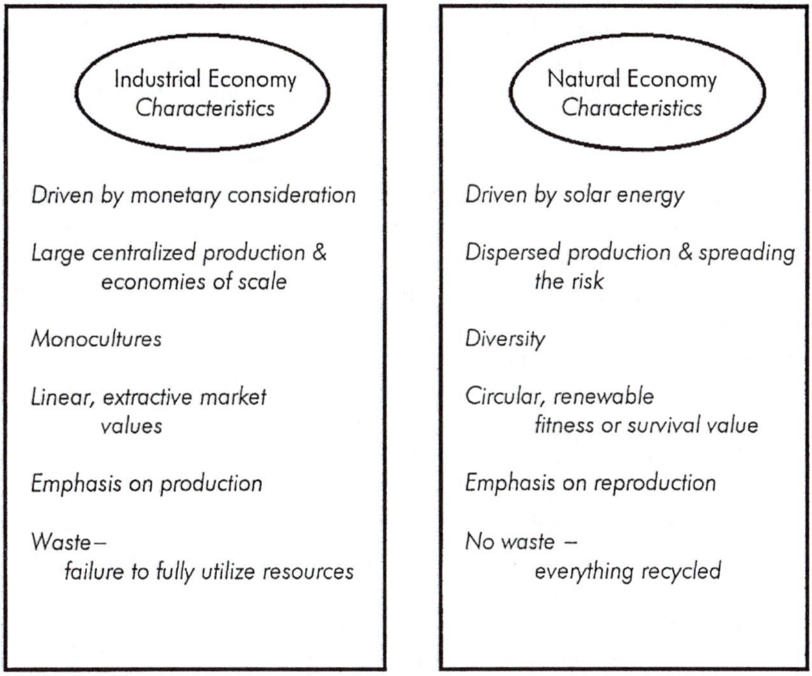

Industrial Economy Characteristics	Natural Economy Characteristics
Driven by monetary consideration	Driven by solar energy
Large centralized production & economies of scale	Dispersed production & spreading the risk
Monocultures	Diversity
Linear, extractive market values	Circular, renewable fitness or survival value
Emphasis on production	Emphasis on reproduction
Waste— failure to fully utilize resources	No waste — everything recycled

Figure 1.1 Characteristics of the industrial and natural economies. (From "A New Vision for Pacific Salmon," Natural Resources Committee, Oregon Business Council, 1996, p. 11. Reprinted with permission of the author.)

documented how the increased gap between rich and poor and the eruption of social violence between Sikhs and Hindus in the Punjab region of India could be traced to the introduction of hybrid seeds — the so-called Green Revolution.

In the quest for sustainable approaches to economic development, questions of how to use science and technology are extremely difficult for two reasons: (1) even with the accumulated social intelligence regarding the effects of past technological innovations,[9] we cannot fully predict or anticipate all the potential effects of any new technology and (2) levels of individual knowledge and differing value frameworks will necessarily spawn debate and disagreement on whether a proposed innovation is likely to have positive or negative effects or whether its benefits outweigh its costs.

From Figure 1.1 one can infer how differing values and technological approaches would permeate an industrial vs. an ecologically grounded economy. Change to an ecological model of science is central and necessary for achievement of a sustainable economic model.[10]

Governance and Public Life

Many contentious theoretical and political questions converge around governance issues and the role of formal governmental entities in the quest for sustainable development. Perhaps not surprisingly, the debate reflects many past and contemporary ideological differences over the proper and effective role of government with respect to social and economic development.[11] For example:

- What kinds of international governance structures will be needed to maintain peace and national security, to protect global environmental commons, to meet human development needs, or to respond to disasters of a magnitude beyond the internal capacity of countries to manage?
- What is the proper and effective distribution of power, planning responsibility, and monitoring and control of resources, among national, regional, and local government units within nations?
- What is the role, the effectiveness, and the democratic representativeness of voluntary organizations in leading sustainability initiatives and making resource decisions? Do traditional or new types of labor unions have a relevant role to play in representing working people in public planning processes?
- How will government units, voluntary non-governmental organizations,[12] and corporate profit-making companies relate to each other in the painful transitional experiments in sustainability? How, practically speaking, should these entities collaborate for sustainability planning? What resources should each contribute? And how should the decision-making structures be designed for community representativeness and checks and balances on excessive political power by any group?

The intense pressure placed on governments, at the 1992 environmental summit in Rio de Janeiro, to agree to improved standards and practices of environmental protection at the global level signifies that many people throughout the world understand and support the need for stronger national and international governance of the global commons (Shabecoff, 1996). The Worldwatch Institute (French, 1995) believes that significant progress has been made on urgent global economic and environmental issues, but that problems grow even more rapidly. Greater democratization of international governance institutions is needed — that is, open meetings and wider representation.

In the past 20 years, with the active advocacy and political support of citizens, many national governments also have made real and significant policy progress on environmental protection. The burgeoning of the global economy (i.e., unfettered trade and movement of capital) during the past two decades has also raised appreciation in some quarters for the important responsibility — and even the ability — of national governments to set environmental standards for industry, monitor work conditions and product safety, and enforce fairness in the social contract between workers and owners of capital (Athanasiou, 1996; Daly, 1996; Daly and Cobb, 1989).[13]

On the other hand, recent studies in environment and sustainable development in the South (Korten, 1995; Rich, 1994) have documented the sometimes nefarious role of national governments in large-scale economic development projects that widen disparities in wealth and poverty and severely damage the physical environment. The collapse of the Soviet Union has permitted exposure of the extreme despoliation of the environment which occurred under totalitarian governments in Eastern Europe (Bolan, 1994; Feshbach and Friendly, 1992). Unhappily, political chaos and unregulated capitalism may be contributing to continued environmental destruction and social collapse in the former Soviet Union.

People's negative views toward government in some parts of the world are based on bitter experience of oppression and fascism (i.e., unchecked, centralized state power) and their close acquaintance with the failure of many government-supported large-scale economic development projects. Among members of some local communities in the United States, deterioration of trust in national government stems from the failure of the Vietnam War, numerous exposés of government corruption, failure of government to solve social problems, and the nation's strongly held cultural heritage of individualism and localism.

These background factors (as well as other socioeconomic forces) have contributed to an emphasis in sustainable development on local initiatives in economic development and environmental renewal and protection. Decentralized, often even voluntary, decision-making structures are touted in much of the literature.[14] Both evaluation research and practice experience suggest that local ownership and participation in planning to solve local problems do maximize "lasting improvement in the quality of community life" (Betancur et al., 1991, p. 199; Gittell, 1990). Representative and responsible governance will require that sustainability initiatives foster dense social relationships as the foundation from which to attack and solve the practical problems of economy and environment.

However, not all political analysts or activists would agree wholeheartedly with the superiority or adequacy of local control. Practically speaking, many environmental and social problems transcend local boundaries and may require a broader level of participation, expertise, and even funding resources for adequate response (Khinduka, 1987; Lélé, 1991, p. 616). For example, in the Pacific Northwest region of North America, planning for salmon recovery requires a regional strategy involving two Canadian provinces and four American states. The cooperation and resources of numerous groups are required: government units, numerous industries, Native people, and other voluntary groups. Roberts (1994) suggests that the sustainability movement may stimulate a revival of regional planning, which in the first half of the twentieth century made a significant contribution to revival of rural regions, including the enhancement of relationships between rural areas and cities.

Another objection to completely voluntary, consensus-based local development planning is that it can become parochial in vision and interest or controlled by local elites. Recently, Michael McCloskey (1996), chair of the Sierra Club, has argued that the voluntary bioregional councils, which are springing up in the United States, can be more easily dominated by industry interests and relieved of adherence to national environmental protection rules. Moreover, as McCloskey points out, urban dwellers also may have a stake in the environmental health of another region, for example, in the management of a national forest, yet be disenfranchised if decision making is relegated to local non-elected stakeholder councils (p. 7). Achievement of sustainable social structures and sustainable patterns of use of environmental resources will require the equitable and representative participation of all sectors and levels of society in decision making, implementation, benefit distribution, and evaluation (Lélé, 1991).

THE PRACTICAL ISSUES IN
SUSTAINABLE DEVELOPMENT PLANNING

Numerous practical issues will form the organizing problems from which to launch sustainability projects. Local communities, whether urban or rural, and national governments must prioritize where to begin, based on their peculiar situation and windows of opportunity. Broadly speaking, sustainable development planning requires attention to a variety of pressing concerns, such as:

- Non-coercive approaches to population control and active promotion of development opportunities for girls and women
- More attention to the family as the basic social unit in which patterns of social interaction are learned and, thus, as the basic cultural unit which must be supported for an economically and environmentally sustainable future for the human family (Garbarino, 1992;[15] Lusk and Hoff, 1994)
- Reduction of unemployment and destitution (through human-scale strategies adapted to local culture and environment)
- Protection of the world's supply of clean air and water, arable land, and food crops
- Reversal of the rapid extinction of species and measures to protect and enhance the biodiversity of local ecosystems
- Development of energy and transportation systems that rely on soft-path, renewable sources and a phase-out of the dependence on fossil and nuclear energy sources
- Redesign of urban areas for sustainability (reducing and reusing waste materials, fostering of urban agriculture and green space, encouragement of denser living patterns, and human-scale building design)

EDUCATION, SKILLS, AND RESOURCES FOR SUSTAINABILITY

Environmental science and economic education will have to be significantly expanded in order to bring about the cultural and economic changes discussed in this chapter and in order to foster wise and active participation in political processes. Popular education and consciousness raising (critical thinking skills) should be a foundational component of every community's sustainability planning.[16] Sustainable development is a concept and a social change methodology that is profoundly humanistic in its emphasis on the human factor in development. Thus, the educational components of initiatives also should incorporate arts, humanities, and social science perspectives.

Practical skills, such as planning, development, and community organization skills, as well as group facilitation and conflict resolution skills are needed and will be learned as groups come together to initiate new directions for their communities. In other words, communities need expertise, but will acquire at least some of it among their own members as they proceed. Scientific knowledge, technical skills, material and technological resources, legal sanction and financial support, and money management skills are essential. In summary, sustainable development is a process derived from a new vision of a society based in humanistic values, democratic politics, respect for the natural world, and a harmonization of wealth-generation goals with human welfare and socio-cultural goals.

REFERENCES

Athanasiou, T. (1996). *Divided planet: The ecology of rich and poor.* Boston: Little, Brown and Co.

Barber, W.J. (1967). *A history of economic thought.* New York: Penguin Books.

Betancur, J.J., Bennett, D.E., and Wright, P.A. (1991). Effective strategies for community economic development. In P.W. Nyden and W. Wiewel (Eds.), *Challenging uneven development: An urban agenda for the 1990s* (pp. 198–223). New Brunswick, NJ: Rutgers University Press.

Bolan, R.S. (1994). Environmental quality in Poland. In M.D. Hoff and J.G. McNutt (Eds.), *The global environmental crisis: Implications for social welfare and social work* (pp. 117–149). Aldershot, England: Avebury Books.

Bothe, M. (1993). The subsidiarity principle. In E. Dommen (Ed.), *Fair principles for sustainable development: Essays on environmental policy and developing countries* (pp. 123–138). Aldershot, England: Edward Elgar Publishing Ltd.

Brecher, J. and Costello, T. (1994). *Global village or global pillage: Economic reconstruction from the bottom up.* Boston: South End Press.

Bruyn, S.T. and Meehan, J. (1987). *Beyond the market and the state: New directions in community development.* Philadelphia: Temple University Press.

Bullard, R. (1990). *Dumping in Dixie: Race, class and environmental quality.* Boulder, CO: Westview Press.

Bullard, R. (Ed.) (1993). *Confronting environmental racism: Voices from the grassroots.* Boston: South End Press.

Carson, R. (1962). *Silent spring.* Greenwich, CT: Fawcett Publications.

Commission for Racial Justice (1987). *Toxic wastes and race in the United States: A national report on the racial and socio-economic characteristics of communities with hazardous waste sites.* New York: United Church of Christ.

Daly, H.E. (1996). *Beyond growth: The economics of sustainable development.* Boston: Beacon Press.

Daly, H.E. and Cobb, J.B., Jr. (1989). *For the common good: Redirecting the economy toward community, the environment and a sustainable future.* Boston: Beacon Press.

Danaher, K. (Ed.) (1994). *50 years is enough: The case against the World Bank and the International Monetary Fund.* Boston: South End Press.

Denniston, D. (1995). *High priorities: Conserving mountain ecosystems and cultures.* Washington, DC: Worldwatch Institute.

Development Alternatives with Women for a New Era (1995). Rethinking social development: DAWN's vision. *World Development, 23*(11), 2001–2004.

Durning, A.B. (1989). *Action at the grassroots: Fighting poverty and environmental decline.* Washington, DC: Worldwatch Institute.

Durning, A. (1992). *How much is enough: The consumer society and the future of the earth.* New York: W.W. Norton & Co.

Feshbach, M. and Friendly, A., Jr. (1992). *Ecocide in the USSR: Health and nature under siege.* New York: HarperCollins.

French, H.F. (1995). *Partnership for the planet: An environmental agenda for the United Nations.* Washington, DC: Worldwatch Institute.

Garbarino, J. (1992). *Toward a sustainable society: An economic, social and environmental agenda for our children's future.* Chicago: The Noble Press.

George, S. (1988). *A fate worse than debt.* New York: Grove Press.

George, S. (1992). *The debt boomerang: How Third World debt harms us all.* London: Pluto Press with the Transnational Institute, Amsterdam.

Gittell, R. (1990). Managing the development process: Community strategies in economic revitalization. *Journal of Policy Analysis and Management, 9*(4), 507–531.

Gunn, C. and Gunn, H.D. (1991). *Reclaiming capital: Democratic initiatives and community development.* Ithica, NY: Cornell University Press.

Hardy, S. and Lloyd, G. (1994). An impossible dream? Sustainable regional economic and environmental development. *Regional Studies, 28*(8), 773–780.

Hoff, M.D. (1994). Environmental foundations of social welfare: Theoretical resources. In M.D. Hoff and J.G. McNutt (Eds.), *The global environmental crisis: Implications for social welfare and social work* (pp. 12–35). Aldershot, England: Ashgate Publishing/ Avebury Books.

Hofrichter, R. (Ed.) (1993). *Toxic struggles: The theory and practice of environmental justice.* Philadelphia: New Society Publishers.

Keynes, J.M. (1964). *The general theory of employment, interest, and money.* New York: Harcourt, Brace & World.

Khinduka, S.K. (1987). Community development: Potentials and limitations. In F.M. Cox, J.L. Erlich, J. Rothman, and J.E. Tropman (Eds.), *Strategies of community organization* (4th ed.), (pp. 353–362). Itasca, IL: F.E. Peacock (orig. published 1969).

Korten, D.C. (1995). *When corporations rule the world.* West Hartford, CT: Kumarian Press and Berrett-Koehler Publishers.

Kunstler, J.H. (1993). *The geography of nowhere: The rise and decline of America's man-made landscape.* New York: Simon & Schuster.

Lélé, S.M. (1991). Sustainable development: A critical review. *World Development, 19*(6), 607–621.

Lusk, M.W. and Hoff, M.D. (1994). Sustainable social development. *Social Development Issues, 16*(3), 20–31.

McCloskey, M. (1996). The skeptic: Collaboration has its limits. *High Country News, 28*(9), 7.

Meadows, D.H., Meadows, D.L., Randers, J., and Behrens, W.W. III. (1974). *The limits to growth: A report for the Club of Rome's project on the predicament of mankind* (2nd ed.). New York: Universe Books.

Merchant, C. (1980). *The death of nature: Women, ecology and the scientific revolution.* New York: HarperCollins.

Midgley, J. (1995). *Social development: The developmental perspective in social welfare.* London: Sage Publications.

Mikesell, R.F. (1992). *Economic development and the environment.* London: Mansell.

Milbrath, L.W. (1989). *Envisioning a sustainable society: Learning our way out.* Albany: State University of New York Press.

Milbrath, L.W. (1996). *Learning to think environmentally while there is still time.* Albany: State University of New York Press.

Oregon Business Council (1996). *A new vision for Pacific salmon.* Portland, OR: Author.

Pearce, D., Barbier, E., and Markandya, A. (1990). *Sustainable development: Economics and environment in the Third World.* Aldershot: Edward Elgar.

President's Council on Sustainable Development (1996). *Sustainable America: A new consensus for prosperity, opportunity, and a healthy environment for the future.* Washington, DC: Author, February.

Rich, B. (1994). *Mortgaging the earth: The World Bank, environmental impoverishment, and the crisis of development.* Boston: Beacon Press.

Roberts, P. (1994). Sustainable regional planning. *Regional Studies, 28*(8), 781–787.

Rogge, M.E. (1994). Environmental injustice: Social welfare and toxic waste. In M.D. Hoff and J.G. McNutt (Eds.), *The global environmental crisis: Implications for social welfare and social work* (pp. 53–74). Aldershot, England: Ashgate/Avebury Books.

Rubin, H.J. and Rubin, I.S. (1992). *Community organizing and development* (2nd ed.). New York: Macmillan.

Sale, K. (1985). *Dwellers in the land: The bioregional vision.* San Francisco: Sierra Club Books.

Schumacher, E.F. (1973). *Small is beautiful: Economics as if people mattered.* New York: Harper and Row.

Sen, G. and Grown, C. (1987). *Development, crises, and alternative visions: Third World women's perspectives.* New York: Monthly Review Press.

Shabecoff, P. (1993). *A fierce green fire: The American environmental movement.* New York: Farrar, Straus and Giroux.

Shabecoff, P. (1996). *A new name for peace: International environmentalism, sustainable development, and democracy.* Hanover, NH: University Press of New England.

Shiva, V. (1991). *The violence of the green revolution.* London: Zed Books.

Taniguchi, C. (1995). Creating an environmentally sustainable city: The Curitiba initiative. *Regional Development Dialogue, 16*(1), 100–107.

Wachtel, P. (1989). *The poverty of affluence.* Philadelphia: New Society Publishers.

Williams, C.C. and Haughton, G. (1994). *Perspectives towards sustainable environmental development.* Aldershot, England: Avebury Books.

World Commission on Environment and Development (1987). *Our common future.* Oxford: Oxford University Press.

ENDNOTES

1. See various reports from the Worldwatch Institute for additional examples of integrated action for environmental protection and development opportunities by grass-roots groups. Worldwatch Paper #88: *Action at the Grassroots: Fighting Poverty and Environmental Decline* (Durning, 1989) and Worldwatch Paper #123: *High Priorities: Conserving Mountain Ecosystems and Cultures* (Denniston, 1995) in particular focus on this topic. See Brecher and Costello (1994) for advocacy research emphasizing the economic dimensions of grass-roots activism.

2. Midgley (1995) cites the Scandinavian countries as the most notable examples of advanced, industrialized countries which have explicitly attempted to develop an integrated approach to economic and social development (p. 51).

3. One creative compromise between Northern environmentalists and peoples of the South is "debt for nature" exchanges. While this strategy preserves a certain portion of the world's natural heritage, it does not address how the local people can sustain their cultural and economic survival. See the writings of Susan George (1988, 1992) for extensive study of the effects of international debt on both lender and debtor countries.

4. See Hilary F. French (1995), *Partnership for the Planet: An Environmental Agenda for the United Nations,* for a brief overview of the "greening" of the Breton Woods institutions; see Kevin Danaher (1994), *50 Years Is Enough: The Case Against the World Bank and the International Monetary Fund,* for a number of social activists' analyses of the social and environmental effects of the policies of the international financial institutions.

5. For the best ongoing chronicle of such experiments in community-based economic development in the United States, see *GEO* (Grassroots Economic Organizing Newsletter, Edina, P.O. Box 213, Willimantic, CT 06226). See the journal *Social Development Issues* for ongoing reporting on community development efforts which incorporate social and economic goals.

6. Sustainability networks can be easily accessed on the Internet.

7. In the following discussion of cultural values, economy, and governance, the author is indebted to *Dwellers in the Land: The Bioregional Vision* by Kirkpatrick Sale (1985). Within the sustainable development movement, bioregionalism focuses on naturally integrated local ecosystems as the proper context for socioeconomic planning and development. The growing focus on planning around watersheds in the western United States represents an application of the bioregional vision. The Willapa Bay case study in Chapter 10 and the Henry's Fork Watershed Council in Chapter 9 are examples of this approach.

8. See *Creating an Environmentally Sustainable City: The Curitiba Initiative* by Cassio Taniguchi (1995) for a description of how this city in Brazil has implemented such a sustainable approach to mass transit.

9. See *The Geography of Nowhere* by James Howard Kunstler (1993) for an incisive, popular tour of the unintended, complex cultural and environmental effects caused by the introduction of the automobile.

10. See *The Death of Nature: Women, Ecology and the Scientific Revolution* by Carolyn Merchant (1980) for a historical analysis, from a feminist perspective, of the values undergirding the development of modern science.

11. See James Midgley's (1995) *Social Development* for a succinct and comprehensive review of the many schools of thought with respect to government's role and actual contributions to modern social and economic development.

12. In United Nations jargon, such groups are usually called NGOs, and thus may include fairly large, non-locally based organizations. In the United States, groups which arise from local initiative and retain local boards of governance are usually called community-based organizations. This serves to distinguish them from other voluntary non-profit organizations, such as large foundations, technical assistance organizations, or large national and international environmental groups.

13. The works cited are important studies of the relationships among environmental concerns and the needs and conditions of workers and communities. Athanasiou (1996) states, "It is a sign of the times that concern about national sovereignty, long a hot button in the South, has become an issue in the United States as well" (p. 173).

14. The preference for local decision making is termed the subsidiarity principle. A detailed analysis of the proper application of this principle to environmental governance was developed by Bothe (1993). He argues that there are no easy answers to the question of the appropriate level of governance: international, national, and local decision making and regulation of environmental practices are needed.

15. Garbarino (1992) is a leading child development expert in the United States. In his book *Toward a Sustainable Society: An Economic, Social and Environmental Agenda for Our Children's Future,* he generates a persuasive and vivid picture of how the mounting threats to the physical environment threaten the promise of a sustainable, high-quality future for children.

16. See the work of Lester Milbrath (1989, 1996) for elaboration of the centrality of learning and education as a foundation for sustainable development. Every sustainability initiative should begin by asking "What do people need to know?" and "What is the best method to educate people about the issue at hand?" Education includes formal and informal instruction, as well as experiential learning from participation and experimentation.

Sustainable Rural
Community Development Cases

CHAPTER 2

Molokai: A Study of Hawaiian Subsistence and Community Sustainability

**Jon Kei Matsuoka, Davianna Pomaika`i McGregor,
and Luciano Minerbi**

Nicknamed the *friendly isle* by the Hawaii Visitors Bureau, Molokai, paradoxically, has been very vigilant in organizing against tourism development. The island, which has the highest percentage of Native Hawaiian residents (of the major islands), is characterized by a relatively arid West End which is owned by a New Zealand–based transnational corporation (Molokai Ranch) that is promoting tourism and attempting to divert water from the island's central aquifer for resort development. At stake are Hawaiian homestead communities that rely on this water for agricultural enterprises and the diversification of the island's economy.

Traditionally, Molokai, with its extensive protected reefs and fish ponds, gained the reputation of a land of *fat fish and kukui nut relish*. Molokai Hawaiians obtained marine resources from the shallow offshore reefs; the deep-sea channels between Molokai and Maui, Oahu, and Lanai (Pailolo, Kaiwi, and Kalohi); the deeper ocean off of the island's north shore; and from an extensive network of Hawaiian-constructed fish ponds.

Over the years, a number of activities contributed to the degradation of the natural environment of Molokai. Offshore reefs and oceans were impacted by pollution, erosion, and soil runoff from tourism, residential development, and ranching. Sand from the West End of Molokai was mined and shipped to Oahu to make cement to build the freeways and hotels and to replace sand loss at Waikiki Beach. Gravel and rocks from East Molokai were used in freeway construction on Oahu. Ranching on the East End contributed to deforestation, erosion, and runoff. Once-productive fish ponds were allowed to fill with silt and the walls fell into disrepair following tsunamis (tidal waves) and storms.

Overharvesting of marine resources is a growing problem. Traditional resources such as the turtle cannot be used for subsistence under new federal regulations. Wildlife such as deer, goats, pigs, and birds are abundant on privately owned lands but are too scarce to be hunted on public lands.

In 1987 the last pineapple company closed its operations. In that same year, a tuberculosis epidemic led to the decision to eradicate all the cattle on Molokai. Molokai General Hospital phased down its operations, stopping all maternity deliveries. Molokai's unemployment rate was three times the state's average at nearly 20% (Department of Business, Economic Development, and Tourism [DBEDT], 1987). Many small businesses shut down. Subsistence fishing, hunting, gathering, and cultivation provided a reliable means of support for the community during the rough economic times.

Many families on Molokai, particularly Hawaiian families, continue to rely upon subsistence fishing, hunting, gathering, or cultivation for a significant portion of their food. Availability of the natural resources needed for subsistence is essential to Molokai households, where the unemployment rate is consistently higher than on other islands and a significant portion of the population depends upon public assistance. In March 1992, the unemployment rate on Molokai was 7.4%, while it was 3.5% statewide (DBEDT, 1992). In March 1993, the unemployment rate of 8.1% on Molokai was still higher than the statewide rate of 4.7% (DBEDT, 1993b). With regard to public assistance, in 1990, 24.4% of the Molokai population received food stamps, 12% received Aid to Families with Dependent Children, and 32.5% received Medicaid. According to the U.S. census for 1990, 21% of the families on Molokai had incomes that fell below the poverty level of $12,674 for a family of four. The ability to supplement meager incomes through subsistence is very important to maintaining the quality of life of families on the island.

Subsistence has also been critical to the persistence of traditional Hawaiian cultural values, customs, and practices. Cultural knowledge, such as about place names, fishing ko`a (shrines), methods of fishing and gathering, or the reproductive cycles of marine and land resources, have been passed down from one generation to the next through training in subsistence skills. The sharing of foods gathered through subsistence activities has continued to reinforce good relations among members of extended families and neighbors.

This chapter is a description of a community-based process that was intended to develop awareness of the plight of subsistence practitioners and the significance of subsistence in terms of cultural and community preservation. In its beginning stages the process centered around a research project. The investigators operationalized a research philosophy best described as *participatory action research* (PAR). This chapter refers to the PAR orientation (methods, results, applications) and provides a general discussion of implications as they relate to efforts directed toward community sustainability.

ORGANIZATION OF THE SUBSISTENCE TASK FORCE

In 1990, after receiving word that Molokai Ranch was in the process of herding wild deer into a 3,000-acre fenced area, the Molokai subsistence hunters picketed the ranch. The hunters organized to stop commercial hunting of the deer by the ranch and sport hunting of the deer under State of Hawaii permits. They also worked with the Nature Conservancy to stop the complete eradication of feral pigs, goats, and deer from pristine rain forest areas with snares. They took their concerns to Governor John Waihee who, after a year of negotiations, decided to form a task force to review all subsistence activities on Molokai.

Governor John Waihee appointed the Molokai Subsistence Task Force in February 1993 to assess the importance of subsistence to Molokai families. The task force was also asked to identify problems related to subsistence fishing, hunting, and gathering on Molokai and to recommend policies and programs to protect and/or enhance subsistence activities on the island.

The task force included representatives of the various sectors of the Molokai community — the Department of Hawaiian Homelands and the Department of Land and Natural Resources, which manages lands for Native Hawaiians and the general public; Molokai Ranch, which is the major owner of private land on the island's West End; the Department of Business and Economic Development and Tourism; *Hui Malama O Mo`omomi,* a group of Hawaiian homesteaders organized to protect the subsistence and cultural resources of the northwest coast of Molokai; and individual subsistence fishermen, hunters, and gatherers. They defined the scope of work and the goals for their study of subsistence on Molokai. To conduct the study, they enlisted the services of a team of University of Hawaii consultants, the authors of this chapter, from the disciplines of social work and health, history and ethnic studies, and architecture and planning.

The process reflected a community-based research and planning effort that involved a multitude of island constituencies, government and private land managers, and social scientists serving as research consultants. The study results were converted into goals and objectives and action strategies which ultimately led to a new marine designation intended to protect natural resources and the opening of access through private ranch lands on the southwest part of Molokai.

RESEARCH PHILOSOPHY AND PARTICIPATORY ACTION RESEARCH

PAR is conducted for the purpose of documenting particular social realities in order to enable constituencies/communities to advocate for communities and the resources needed to sustain them (Whyte et al., 1991). A basic premise stemming from this model is that research can be a catalyst for social action. PAR

can be viewed as a process that involves establishing networks between representatives of a locale or community, research consultants, and bureaucrats. The
goal of this process is to discover and document aspects of a community in order
to preserve or ameliorate such aspects in an effort to sustain or improve the quality
of life. Program and policy implications are drawn from the data, and social
action and implementation strategies are developed by community planners.

The role of the researcher needs to be redefined in the context of working
with indigenous groups. The conventional model of pure research in which
participants are treated as passive subjects and kept at arm's length throughout
the technical and interpretive stages of research is antithetic to the goal of
documenting indigenous realities. According to Whyte et al. (1991), the greatest
conceptual and methodological challenges come from engagement with the
world. Thus, a prerequisite to PAR for any researcher is to be able to connect
on a personal level with members of the community under study. Developing an
appreciation for and sensitivity to the subjective realities of residents of a locale
is essential to humanizing or indigenizing scientific inquiry. Engagement also
refers to the ability of the researcher to elicit and incorporate indigenous concepts into measures and methods, while maintaining a high level of scientific
rigor and objectivity. The results must be able to stand up to the scrutiny of
hostile reviewers.

The researchers for this study represented a variety of disciplines. The
interdisciplinary team approach was useful in promoting and integrating various
orientations and methodologies. A multimethod research approach is consistent
with an ecological orientation and the goal of assembling multiple social dimensions into a mosaic form.

Research Methods for the Molokai Subsistence Study

The questions and format of the survey were developed by the task force over
the course of several meetings. The survey was designed to provide an assessment of subsistence activities as practiced and experienced by all of the people
living on Molokai.[1] Extensive input from task force members enabled the
university consultants to develop a questionnaire that was relevant, comprehensive, objective, and culturally appropriate. The survey instrument was designed
to acquire data in the following areas: (1) demographic characteristics of respondents and rates and types of subsistence practices, (2) the significance of
subsistence to one's family (e.g., the percent of a family's food that comes from
subsistence), (3) the types of non-consumptive uses of subsistence resources
(e.g., sharing, exchange), (4) the cultural significance of subsistence, and (5) the
types of issues and problems that impede subsistence on Molokai.

A telephone survey was conducted to acquire a random sample of Molokai
residents. The response rate among those contacted was 89%. In order to facilitate the research process and on-site organization, two Molokai-based coordinators from a community development project were hired. A total of six Molokai

residents were hired and trained to conduct the interviews. The interviewers underwent several hours of orientation and interviewer training.

The qualitative component of the study consisted of focus group discussions with subsistence practitioners from each of the districts on the island. Eight focus group meetings, involving a total of 105 participants, were held within a one-month period. Prospective participants were contacted by the task force to attend the gathering in their district in order to share information on their customary practices and concerns regarding subsistence resources. At their request, a focus group session was also held for commercial fishermen.

Another component of the study that occurred in the focus group sessions was the mapping of subsistence sites and areas. Participants were provided colored-coded stickers that corresponded with different types of subsistence practices. They were asked to place the stickers on a topographic map in order to locate areas where they engaged in subsistence. The dot locations were digitized into a Geographic Information System (GIS) map.

The respondents from the telephone survey were asked to identify (by name) areas where they engaged in subsistence activities. Maps describing the percentage of use by localities were then constructed. This information was compared with the dot maps generated from the focus groups by the Hawaiian subsistence practitioners.

In this study, the multimethod approach was useful in addressing the particular informational needs of the task force. The survey provided a general scope of subsistence practices for the entire island and was used to document the nature and significance of these practices. The focus groups were a means of collecting more specific data on the issues and problems associated with subsistence. The GIS maps served to expand the database and provided critical information regarding the location of significant subsistence area.

Survey Results

The survey questionnaire was designed to collect data in a variety of areas related to subsistence. Demographic data were collected to determine the representativeness of the sample and to conduct subgroup comparisons on aspects of subsistence. In terms of the 1990 U.S. census, the survey sample was representative of the island's population in terms of ethnicity, age, gender, districts of residence, and overall demographic composition. Other demographic variables that were critical to the analyses included the place where respondents grew up (or spent most of their childhood), length of residency, and household size.

The following is a summary of the major results from the survey:

- Respondents were asked to estimate the number of times per month other Molokai residents gave their family food like fish, meat (deer, pig, goat), or limu (seaweed) that they caught, gathered, or grew themselves. The average number reported was 4.3 times per month. This averages about once per week.

- The overall importance of subsistence to families on Molokai was assessed by asking respondents to rate its importance on a four-point scale. The average or mean score was 1.9, which fell within the *important* range.

- In order to assess the significance of subsistence as a source of food for Molokai families, respondents were asked to estimate the total percentage of their food that came from various subsistence activities.[2] The average estimated amount of food was 28%. Twenty-five percent of the respondents (54) stated that 50% or more of their food came from subsistence activities. The Native Hawaiian subgroup reported that 38% of their food was derived from subsistence (most of any ethnic group).

- Respondents were asked to describe ways in which resources derived from subsistence were used (other than for consumptive purposes). *Sharing* and *gift-giving* were the most common purpose. The second highest response was *exchange* and *trade*. The use of resources derived from subsistence for sale or commercial purposes was reported by a small number of respondents.

- The social, cultural, and health benefits of subsistence were also examined. The categories that received the highest number of responses were *exercise/health/diet, family togetherness,* and *recreation*.

- Respondents were asked whether the resources derived were used for occasions other than usual family consumption. Seventy-two percent (175) stated that they used subsistence resources for special occasions. *Birthdays* and *luaus* were the special occasions for which respondents collected resources most often.

- Respondents were asked to rate issues/problems that may have served to impede their ability to subsist. The problems cited most often were *pollution, waste of resources, erosion, overdevelopment, people who take too much,* and *lack of access*.

- Using a four-point scale, respondents were asked to rate the importance of subsistence activities in regard to the lifestyle of Molokai. The mean response fell within the very important range, indicating that respondents believed that subsistence was very important to sustaining Molokai's lifestyle.

Statistical analyses were conducted in order to examine differences between demographic subgroups on items related to subsistence. The information derived from these analyses was useful in determining the meaning and importance of subsistence to various communities or constituencies on Molokai. Analysis of variance (ANOVA) statistical tests were used to determine differences between subgroups on numerous items. The aspects of subsistence (dependent variables) that were examined included sharing, exchange, sale, carrying on the culture, family togetherness, spirituality, exercise, recreation, medicine, learning, crafts, special occasions, and rates of the different types of subsistence. Rather than presenting all of the ANOVA results in tabular form, the authors will provide a synopsis of the results that were statistically significant (at least 0.005 or lower).

- Significant differences were found on 13 of the analyses (out of 16) comparing age subgroups. In every instance, younger people (18 to 39) were more involved in subsistence activities and placed a higher value on aspects related to subsistence compared to the oldest cohort (60 and older).
- Differences were found between men and women on four of the analyses. Men engaged in subsistence activities (fishing, hunting, ocean gathering, raising livestock) significantly more than women.
- In terms of ethnicity, Hawaiians were more involved in subsistence activities and placed a higher value on aspects of subsistence than the other ethnic groups on all 16 analyses.
- On the variable representing where people grew up or spent most of their childhood, those who grew up on Molokai were significantly more involved in subsistence activities and placed a higher value on aspects related to subsistence compared to those who grew up elsewhere. Significant differences were found on 14 of the analyses. Similarly, long-time Molokai residents (20 years or more) were involved in subsistence activities (8 analyses) compared to more recent residents (1 to 9 years).
- Household size accounted for significant differences in subsistence outcomes. Larger families (7 to 12 members) placed a higher importance on and engaged in more subsistence activities compared to smaller families (1 to 2 members). Significant differences were found on 14 of the analyses. Larger family size may also be correlated with being Hawaiian as well as a consequence of being younger. Older people tend to have fewer children in their households.

LOCATION OF SUBSISTENCE SITES USING GIS MAPS

This integrated information approach provided important data on the location of subsistence areas. Areas were identified in terms of the type of subsistence activity that occurred there (Figure 2.1). The presence of a particular resource could also be extrapolated from the data. The GIS maps were a useful planning tool because they provided a graphic depiction of areas throughout the island of Molokai that were critical to the perpetuation of subsistence activities. The map also provided insights into which areas might require protection from overuse.

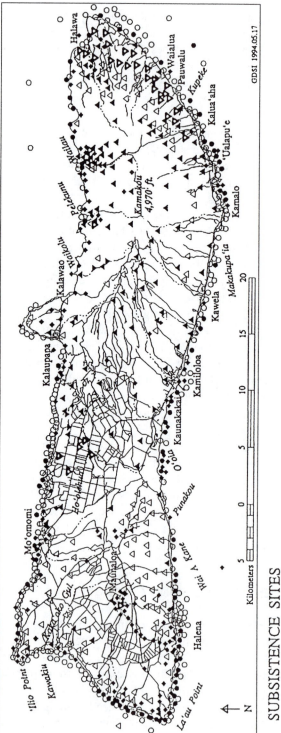

Figure 2.1   Subsistence sites. (From Focus Group Participants Molokai Subsistence Task Force Report, 1994.)

Focus group participants generally identified sites that were within their own district of residence. This information provided clues regarding the proximal relations between residents of a particular locale and resource sites. Activity and land use patterns could be conceptualized in terms of a range within which tenants of a locale were allowed to practice subsistence. The range was determined by factors such as accessibility, convenience, knowledge of and familiarity with an area, ancestral ties, and proscriptive norms which discouraged one from venturing into someone else's range.

DISCUSSION OF THE SOCIAL AND ECONOMIC BENEFITS OF SUBSISTENCE ACTIVITY

The prevalence of subsistence on Molokai was reflected in the amount of food derived from these practices and feelings about its overall importance to families. The fact that families were highly dependent on subsistence for survival, especially Hawaiians, pointed to the value of subsistence as a sector of the economy. This dependency on subsistence resources was even more paramount when examined against the backdrop of relatively low income levels on Molokai. Close to half of the sample made less than $20,000 annually. This low monetary amount had implications for purchasing power, diet, recreation, and family and community dynamics.

Without subsistence as a major means for providing food, Molokai families would be in a dubious situation. Subsistence provided families with the essential resources that compensated for low income and a means of obtaining food items that may have been prohibitively costly under a strict cash economy. Food items like fish, limu, and deer meat, which were normally obtained through subsistence, are generally unavailable or are very costly in stores. If families on fixed incomes were required to purchase these items, they would probably opt for cheaper, less healthy foods, which would predispose them to disease and other health problems. In this respect, subsistence not only provided food but also ensured a healthy diet which was critical to the prevention of disease.

On a related issue, subsistence generally requires a great amount of physical exertion (e.g., fishing, diving, hunting) that burns calories and improves aerobic functioning. It provides a valuable form of exercise and stress reduction that contributes to positive physical and mental health. Subsistence also requires a lot of time. Those who engage regularly in subsistence were less prone to the types of problems that afflicted those who were at a loss for meaningful activities. The lack of activities is often associated with lethargy, boredom, or other conditions that contribute to obesity, substance abuse, and so forth.

According to the results of the study, subsistence is analogous to recreation for a majority of respondents. It is a form of recreation that, once all of the essential equipment has been obtained or made (e.g., fishing tackle, diving gear), is relatively inexpensive. Unlike most other forms of recreation, which are costly every time they are engaged in (e.g., golf greens fees) and intended to provide a sense of psychological fulfillment, subsistence has economic and cultural benefits as well.

Beyond the immediate economic and health advantages that come with subsistence are other qualities that serve to enhance family and community cohesion and perpetuate culture and spirituality. Subsistence is an activity that provides prescribed roles for its members. Family members of all ages feel that they contribute to family welfare through their involvement in subsistence. Subsistence activities are a central part of camping trips or family outings, and parents and children alike are involved in catching fish and gathering marine resources. Older children are oriented toward subsistence by their elders who teach them about techniques and the behaviors of various species.

On another level, subsistence provides a basis for sharing and gift-giving within the community. Residents generally subscribe to a process of reciprocity and sharing with those who are unable to obtain resources on their own. Families and neighbors exchange resources when they are abundant and available, and the elderly are often the beneficiaries of resources shared by younger, more able-bodied practitioners. Some practitioners believe that they must share their catch with others even when it is meager, because generosity is rewarded by better luck in the future.

Resources obtained through subsistence are used for a variety of special occasions that bond families and communities. Resources such as fish, *limu* (seaweed), `*opihi* (limpid), deer meat, etc. are foods served at birthdays, luaus, graduations, and holiday celebrations. `*Ohana* (family members) and community residents participate in these affairs that cultivate a sense of communal identity and enhance social networks.

Time spent in nature cultivates a strong sense of environmental kinship that is the foundation of Hawaiian spirituality. Subsistence practitioners commune with Nature, honor the deities that represent natural elements and life forces, learn how to *malama* (take care of) the land, and develop an understanding about patterns and habits of flora and fauna.

While traversing the land, practitioners also become knowledgeable about the landscape, place names and meanings, ancient sites, and areas where rare and endangered species of flora and fauna exist. This knowledge is critical to the preservation of natural and cultural landscapes because they provide the critical link between the past and the present. For example, *wahi pana* (sacred sites) that are referred to in ancient chants and legends are often lost amidst changes due to modernization. The identification or rediscovery of these sites provides a continuity that is critical to the survival and perpetuation of Hawaiian culture.

An inherent aspect of traditional subsistence is the practice of conservation. Traditional subsistence practitioners are governed by particular codes of conduct that are intended to ensure the future availability of natural resources. Rules that guide behavior are often tied to spiritual beliefs concerning respect for the `aina (sacred land), the virtues of sharing and not taking too much, and a holistic perspective of organisms and ecosystems that emphasizes balance and coexistence.

The finding that younger cohorts were more involved in subsistence and related practices than older people is not surprising given that the former group is more physically active and generally has more dependents to feed and care for. This finding may also reflect a resurgence of or renewed interest in traditional Hawaiian practices among younger people. Men were more involved in various types of subsistence than women. This result reflects gender role variations for particular subsistence activities. Traditionally, activities such as fishing and hunting were done by men, while women engaged in plant gathering. The fact that men continue to dominate these activities points to the continuation of certain traditions.

Hawaiians engaged in subsistence and related practices more than other ethnic groups. This finding reflects the importance of subsistence to this group and the perpetuation of culture through subsistence activities. As mentioned previously, subsistence also plays an important economic role, and this may be especially true for Hawaiians who generally have lower incomes. The fact that Hawaiians engage more in subsistence than others also provides evidence that these activities are embedded in the culture and can be explained through a history of adaptation, the development of an indigenous economy, and the maintenance of cultural traditions despite the influx of foreign lifeways. It is important to note that the other groups (e.g., Filipinos, Japanese) engaged in subsistence, although not at the same level as Hawaiians.

Those raised on Molokai had higher rates of subsistence and related activities than those from other places. This can be explained by the unique subculture of Molokai that is manifested through its lifestyle and socialization practices that encourage subsistence. Those raised elsewhere are not exposed to the same socialization experiences, especially if they come from urban environments on the American continent and elsewhere. Subsistence may not be a part of their growing up because it was not stressed within their culture and resources were not available. The same process held true for long-time residents. Whether a function of age, generation, or exposure over time, the longer one lives on Molokai, the more likely one is to engage in subsistence.

Finally, large families (households) engaged in subsistence more than single people or those with smaller families. This again points to the economic benefits derived from subsistence, especially in family situations where there are many people to feed. Larger families or `ohana may also possess more traditional values than smaller families because they reflect a traditional structure comprised of multiple generations. Thus, they are more inclined to engage in subsistence. Smaller families tend to be nuclear, reflecting a physical separation from parents or grandparents, who are a crucial element to the perpetuation of cultural values. Smaller families may also be comprised of older members whose children have migrated to other locations.

TRENDS AND ISSUES IDENTIFIED IN FOCUS GROUPS

Focus group discussions reinforced the points that subsistence is vital to the economic, cultural, and social well-being of Molokai families and that it is widespread and actively practiced. There is a growing concern on the island that mounting pressures are leading to overharvesting, which will ultimately wipe out the natural resources upon which the community relies for sustenance. At the heart of the matter is recognition of, and conformance to, traditional Hawaiian subsistence values, customs, methods, and practices. The primary reason why Molokai has the natural resources it needs for subsistence is because previous generations of subsistence practitioners lived in accordance with `ohana values of sharing and respect and faithfully followed traditional and customary practices and *kapu* (rules of conduct).

The present generation of subsistence practitioners is facing new challenges and problems from tourism, commercialism, and newcomers who are ignorant of Hawaiian subsistence values, customs, and practices. Hawaiian practices that are customarily passed down from one generation to the next are in decline because of increasing competition from off-island fishermen and hunters and new residents from the North American continent and the Philippines. There is a growing sentiment that *if you don't take something when you see it, someone else will.* Thus, rather than taking only what is needed, more is harvested and sometimes wasted. The widespread use of large freezers also contributes to overharvesting. In the past, the ocean was the *icebox* and one gathered only enough for the `ohana and close neighbors and *kupuna* (elders) to eat. More recently, subsistence practitioners gather more than what their families can immediately consume and the surplus is stored in freezers.

Many of those who have not been trained by *kupuna* in subsistence skills are using improper methods to harvest. For example, *limu* beds are disappearing because people are pulling it up from its roots rather than plucking it.[3] The traditional Hawaiian practice which dictated that only mature resources be gathered and that the reproductive cycles be respected is not honored by newcomers. Thus, juvenile marine life is being harvested. Fish, squid, and lobster are being harvested during their spawning season, when they congregate near the shore and are easier to catch. Gill nets and lobster nets are indiscriminately trapping marine life, and some areas are showing signs of overfishing. Among deer hunters, the common mentality is to go after the trophy rather than to get food for family and neighbors. Subsistence hunters believe that this practice serves to reduce the herd count. Night poaching of deer, which is becoming more common, poses a danger to public safety and contributes to the wasting of meat. Soaring prices for `opihi in markets and catering businesses on Oahu, where the limpid have been wiped out, are leading to increased harvesting for commercial sale.

RECOMMENDATIONS FROM FOCUS GROUPS

The natural resources of Molokai and its surrounding waters are still considered sufficient to support both subsistence and commercial harvesting; otherwise, subsistence practices would not be as widespread. However, the resources are not as abundant as adult subsistence practitioners remember them to be when they were growing up. Molokai subsistence practitioners have arrived at a crucial juncture. There is increasing concern that if something is not done to reverse the trend of overharvesting and diminishing resources, there will be nothing left for future generations. Participants felt that the key to restoring a balance between subsistence harvesting and diminishing natural resources is a community-wide acceptance of traditional Hawaiian subsistence values and practices. These notions need to be promoted, understood, accepted, and practiced by everyone who engages in subsistence on Molokai, no matter what their ethnic background.

According to the participants, there is a need for a commitment by everyone in the community to manage the natural resources of Molokai — not just to benefit the current generation but for the well-being of future generations. This commitment can be secured primarily through educational programs to provide training in proper methods of harvesting resources and to inspire acceptance of the traditional values of caring for, and nurturing, the land and the ocean. Education should occur at all levels, including the Molokai school system, Department of Land and Natural Resources education initiatives such as hunter education classes, and community organizations disseminating information about resource management.

New fishing rules and regulations and community-based management of natural resources will also be a way to curb current trends in overharvesting. The Department of Land and Natural Resources will need to increase the number of enforcement officers assigned to Molokai. However, government enforcement is not seen as a solution to better management of the island's resources. Subsistence and commercial users need to take responsibility for their own actions. Volunteers, peer pressure, and community-based resource managers can more effectively promote the proper utilization of resources.

Restocking will also be an important component of sustaining subsistence resources on the island. Natural hatcheries, such as at Mo`omomi Bay, need to be protected as fish sanctuaries. This area is seen as critical to fish breeding. The Department of Land and Natural Resources should streamline the permitting process for community-based economic development groups to reopen the traditional fish ponds which are now part of the ceded public lands trust.[4] Hatchery programs should be developed to foster the propagation of marine life in the fish ponds and in selected bays around Molokai.

The other major area of concern to subsistence practitioners on the island is the provision of customary access to all parts of the island. Of particular concern are areas of Molokai Ranch that were formerly open under the pineapple company but have since been closed by the new landowners. It would be acceptable to have access regulated by the use of permits and keys. Limiting access to certain areas of the island to foot trails would also serve to limit the amount of resources which can be harvested.[5] A relationship of mutual trust and management responsibility should be established between subsistence practitioners and Molokai Ranch.

In summary, subsistence on Molokai will continue to be essential to the lifestyle of the people. Community-based management of the resources, rooted in the traditional values of *aloha `aina* (love of the land) and *malama `aina* (taking care of the land), is vital to sustaining a subsistence lifestyle. Another critical factor in sustaining subsistence activities and the protection of natural resources will be the recognition of subsistence as an essential and viable sector of the island's overall economy.

SUBSISTENCE AS A SUSTAINABLE SECTOR OF MOLOKAI'S ECONOMY

A primary reason for the persistence of subsistence practices on Molokai has been the continued availability of renewable natural resources. In turn, while years of macroeconomic strategies have wreaked havoc on Hawaii's natural environment and endemic species of flora and fauna in urban areas and on plantations, subsistence practices have allowed the natural resources in rural communities such as on Molokai to persist.

Despite the resilience of subsistence on Molokai up to this point, a key concern among focus group participants was how long subsistence practices could be maintained in the face of diminishing returns. Unless drastic and decisive measures are undertaken to protect habitats and the critical mass of species required for regeneration, future generations may not be able to engage in subsistence practices for lack of adequate returns. That is, the amount of resources obtained will not be worth the amount of effort exerted.

A key dimension to the theory of sustainable development is how to offset environmental degradation through preservation. This dimension is germane to our understanding of the issues that surround the Molokai Subsistence Task Force. Although Molokai's population has remained static over time, burgeoning neighboring island populations have resulted in intense competition over resources that are considered to be rightfully those of Molokai residents. Because of overharvesting and resource depletion in places like Oahu and Maui, subsistence and commercial harvesters have sought to exploit the more abundant resources of Molokai. Problems have occurred because of conflicting views about territoriality and tenants' rights, perceived threats to Hawaiian traditions by greedy users who take too much, more efficient technologies (e.g., faster boats) that have overwhelmed natural carrying capacities, and other influences.

The most common concern among those who are identified as traditional practitioners is that current trends will impair the future productive capabilities of the `aina. The natural equilibrium based on rates of *take* and replenishment has been disturbed by heightened competition over resources and environmental degradation. This seriously reduces the opportunity for future generations to partake in the traditional activities that are believed to be at the basis of Hawaiian well-being.

Beyond the direct resource and material rewards resulting from a subsistence economy are cultural benefits that are critical to community and family well-being. A subsistence economy emphasizes sharing and redistribution of resources, which creates a social environment that cultivates community and kinship ties, emotional interdependency and support, prescribed roles for youth, and care for the elderly. Emphasis is placed on social stability rather than individual efforts aimed at income-generating activities. We found that large families were more dependent than smaller families on subsistence resources, and all members who were old enough played a role in gathering resources. When a resource was caught or gathered in large quantities during certain seasons, it was common practice to share with `ohana or community members. The *kupuna* or elderly were especially reliant upon the process of sharing and exchange because many were not able to engage in strenuous physical activities associated with subsistence. In their earlier years, they were benefactors in this same process. Subsistence, as a process of sustainable development, is a value-laden economic system that places emphasis on social relations over exponential growth rates (Halapua, 1993).

Given all of these factors, subsistence has been a viable sector of the economy that has continued to function alongside the sugar and pineapple plantations and the ranches. Hawaiian extended families commonly supplemented their incomes with subsistence fishing and hunting. Unfortunately, subsistence is generally not recognized as a bona fide economic sector by Western economists. In the face of economic decline in Hawaii (such as with the phasing out of agribusiness), decisions are generally made that promote new economic development based on a linear process toward capital accumulation (Department of Urban and Regional Planning, 1989). This usually comes in the form of tourism.

Molokai provides a rare example of how residents adapted to changing economic circumstances without massive external intervention. Historical accounts have indicated that when agribusiness closed on Molokai, subsistence became a more vital aspect of the economy. Through community-based efforts, residents organized to successfully stave off tourism development while promoting values related to community and family integrity. Subsistence and other community-based endeavors are considered the forces that bind together the social elements necessary for cultural perpetuation. Subsistence should not be viewed as a replacement economy per se, but as a tradition that has survived after macroeconomic strategies (i.e., plantations, ranches) failed.

The impact of tourism and related commercial activities on subsistence is not seriously factored in as an economic or social cost. In other words, subsistence is usually not assessed in terms of how it will be impacted or considered as a viable alternative that will at least partially compensate for the loss of jobs and revenues in a declining economy. The most common trend supported by government and labor unions is to find quick replacements to plantation closings. Thus, little is known about how communities fare when left to their own devices in the aftermath of a failed economy. What is not taken into account in the decision-making process is people's *staying power* or their commitment to a place where they often have genealogical ties, cultural heritage, and their willingness to try alternative approaches to achieving sustainability.

Whatever economic recovery strategy is selected, it should allow for subsistence to continue to play a significant role. This is especially critical on Molokai, where natural resources are available and subsistence is an integral part of lifestyle. Community planning is a proactive strategy that should encourage a functional coexistence and balance among subsistence, the market economy, and government.

ACTION PLAN

The data derived from the study served as a basis for generating a plan of action intended to address the issues/problems identified by participants. The task force, through lengthy discussions, developed strategies that were deemed both realistic and effective. The action plan consisted of six major components:

- Establish the Mo`omomi Subsistence Fishing Area
- Establish ongoing negotiations between subsistence practitioners and Molokai Ranch regarding access
- Develop educational programs for Molokai residents
- Endorse a homesteader management plan for Hawaiian Homeland hunting grounds
- Amend the Hawaii Fishing Regulations
- Appoint a Molokai Subsistence Advisory Committee

As of January 1997, the first four components were undertaken and implemented. The remaining two have not yet been pursued.

Mo`omomi Subsistence Fishing Area

Through the efforts of the Molokai Subsistence Task Force and *Hui Malama O Mo`omomi,*[6] the Seventeenth Legislature (1994) passed a law (Act 271/94) to establish *community-based subsistence fishing areas* throughout the islands.

Under the law, the Mo`omomi area was set up as a demonstration pilot project. *Hui Malama O Mo`omomi* and the Department of Land and Natural Resources worked together to establish the Mo`omomi Subsistence Fishing Area as a demonstration project. The bays of Mo`omomi and Kawa`aloa from Na`aukahihi point in the east to Kai`ehu point in the west were recognized as the community-based subsistence fishing area. Within this area, commercial fishing and certain fishing methods (e.g., laying net) are restricted and permits are required to engage in subsistence fishing.

Molokai Ranch Access

The task force asked Molokai Ranch to recognize traditional Hawaiian rights of access based upon established custom and tradition. A letter was sent to the ranch and discussions with ranch representatives and attorneys were initiated. The Department of Land and Natural Resources reminded the ranch that the provision of access through its West Molokai lands was a condition of a conditional use permit at the time that it had been granted in the 1970s. Despite the efforts of the task force, the ranch failed to open access.

In 1996, the ranch sought to renew its lease of the Hale O Lono Harbor along the south shore of West Molokai. Two Hawaiian subsistence practitioners intervened in the permitting process and sought to open access to Hale O Lono Harbor. In response, the Department of Land and Natural Resources required that Molokai Ranch provide access to Hale O Lono Harbor as a condition for renewal of the lease agreement. In November 1996, Molokai Ranch announced that limited access would be afforded on two weekends each month through January 1997. Access will continue through 1997 on terms that the community and the ranch will negotiate.

Education

Hui Malama O Mo`omomi has an ongoing program of education for students in the Molokai schools. The fishermen and gatherers demonstrate to students how to properly harvest the ocean resources and protect Molokai's fragile natural resources. They have been featured in educational video productions about Molokai which have aired on Hawaii public television. Funds have been obtained to produce an educational video on the Mo`omomi Subsistence Fishing Management Area.

Hunting

The Department of Hawaiian Homelands has turned over management of game on the Molokai Hawaiian Homelands to the homesteaders in Ho`olehua on the West End of Molokai.

LESSONS FOR RESEARCHERS AND FOR OTHER COMMUNITIES

Many lessons have been learned from the Molokai Subsistence Task Force experience that may be of use to other rural and indigenous communities attempting to protect their resources and lifestyle. The efforts by the task force are replicable when properly adapted to community-specific conditions. Some of the lessons learned and recommendations to other communities are as follows:

- Understand and appreciate the lifeways of a community. The organic sociocultural system that maintains community equilibrium is often based on a delicate ecological balance. There are fundamental qualities that sustain a rural, indigenous community that require protection from misuse and development.
- The will or desire to sustain a community and its vital resources is an inalienable human right that often transcends Western law. It is essential to staying healthy and maintaining a preferred quality of life.
- Communities must persevere and must not be intimidated by Western law or values. As in the case of Molokai, a great deal evolved from the efforts of practitioners who expressed their concerns. Culture is to be cherished and practiced. Indigenous rights must be exercised or else they will vanish.
- Community organizers and grass-roots organizations such as in the case of *Hui Malama O Mo`omomi* need to connect with resource-controlling and decision-making institutions (e.g., non-government organizations, state/federal agencies, universities). Establishing working relationships with various groups invokes greater leverage and provides organizers with insights into the political process. In the end, institutional representatives may advocate for proposals put forth by grass-roots organizations.
- The process of data discovery and community-based planning should be inclusive of all sectors of the community. Input from a variety of constituencies is essential to community assessments. Inclusion also serves to dispel suspicion and to prevent special interests from dominating the decision-making process.
- Results from scientific inquiry need to be translated into operational or action terms so that strategies and objectives can be clearly articulated and thus understood by government bureaucracies.
- Propose or operationalize recommendations as pilot demonstrations. Implementation of programs or policies is a first step toward public and government approval.
- Communities need to monitor government rules and regulations. Authorized agencies may not have the resources to effectively manage lands and human behavior — particularly when powerful landowners/ stakeholders are involved. Communities may be empowered to manage their own domains.

REFERENCES

Brewer, J. and Hunter, A. (1989). *Multimethods research: A synthesis of styles.* Newbury Park, CA: Sage Publications.

Department of Business, Economic Development, and Tourism (1987). *The state of Hawai`i data book.* Honolulu, HI: Author.

Department of Business, Economic Development, and Tourism (1992). *The state of Hawai`i data book.* Honolulu, HI: Author.

Department of Business, Economic Development, and Tourism (1993a). *Report of the Governor's Task Force on Moloka`i Fishpond Restoration.* Kaunakakai, HI: Author.

Department of Business, Economic Development, and Tourism (1993b). *The state of Hawai`i data book.* Honolulu, HI: Author.

Department of Urban and Regional Planning, University of Hawai`i (1989). *Sustainable development or suburbanization? Cumulative project impacts in Ewa and Central O`ahu.* Manoa: University of Hawaii.

Halapua, S. (1993). *Sustainable development: From ideal to reality in the Pacific Islands.* Paper presented at the Fourth Pacific Islands Conference of Leaders, Tahiti, French Polynesia, June 24 to 26.

U.S. Bureau of the Census (1990). *Census of the population: Characteristics of the population.* Washington, DC: Author.

Whyte, W.F., Greenwood, D.J., and Lazes, P. (1991). Participatory action research: Through practice to science in social research. In W.P. Whyte (Ed.), *Participatory action research* (pp. 19–55). Newbury Park, CA: Sage Publications.

ENDNOTES

1. The definition of subsistence used in the study was the customary and traditional use by Molokai residents of wild and cultivated renewable resources for direct personal or family consumption as food, shelter, fuel, clothing, tools, transportation, culture, religion, and medicine and for barter, or sharing, for personal or family consumption and for customary trade.

2. Overall estimates of the total percentage of food derived from subsistence were calculated from weekly estimates. Seasonal variations were accounted for in the findings.

3. Plucking *limu* or tearing it at the base of the plant allows for regeneration. Pulling it out from the roots essentially kills the plant.

4. Ceded lands will be returned to the Hawaiian Nation once it is established. There are current state-funded initiatives and independent efforts to restore a sovereign Hawaiian nation.

5. Having to transport resources by foot limits the amount of resources one can harvest. Thus, areas that restrict vehicular travel serve to protect areas from overharvesting.

6. *Hui Malama O Mo`omomi* is an organization of Hawaiian homesteaders living in Ho`olehua on the island of Molokai. After the work was completed and the task force disbanded, *Hui Malama O Mo`omomi* was the remaining entity that pursued the recommendations.

CHAPTER 3

Community Rebuilding in the Philippines: A Poverty Alleviation Program in Negros Occidental, 1990–1995

David R. Cox

The Philippines is an archipelago made up of over 7,000 islands. Of these, however, eleven constitute 94% of the land area; one of these is Negros, one of the Visayas group in central Philippines. These islands of the Philippines are among the most spectacularly beautiful in the world. Most, like Negros, have rugged interior uplands, with volcanic peaks, potentially lush coastal plains, and surrounding shallow waters.

Negros Occidental, one of the two provinces which make up Negros, is the setting for this chapter. It has comparatively level coastal plains in the west and north and an upland mountain range with peaks rising to 8,100 feet. Six large rivers carry the high rainfall of the rainy season down across the plains, and their surrounding areas are flood-prone during the southwest monsoons of the rainy season. The coastal plains were cleared for sugar plantations in the eighteenth century, and much of the uplands was logged in recent decades. By 1990, forest land in the province was down to 33% of the land area, and actual forests were a mere 5%.

The background of Negros Occidental is complex, reflecting the complex history of the Philippines, and that complex history is relevant to what is happening in Negros Occidental today. As in most of the country, the people of Negros Occidental had traditionally organized themselves into barangay (smallest local government area), with a spirit of mutual collaboration which has aided the most recent efforts toward community development and community-based poverty alleviation. The Spanish invasion of 1565, however, brought many changes to this traditional pattern, and an economic boom in the latter half of the eighteenth century, based on the sugar industry, resulted in large-scale in-migration. In Negros Occidental, the areas open to cultivation were formed into haciendas (estates) and devoted to sugarcane production, which imposed a set of feudal-type relationships. In 1973–82, the sugar industry employed 535,000 workers, with 70% of the cultivated area devoted to it and a sugar output of half the national total.

Then, in the mid-1980s, recession hit the sugar industry, and the basis of the livelihood of hundreds of thousands of people disappeared virtually overnight. This occurred in the context of a shrinking Philippine economy, shrinking by 5% in both 1984 and 1985. Poverty then afflicted the lives of most of these people, as they tended to have no land or an alternative source of food supply.

The government of the Philippines acted quickly. This was in the immediate aftermath of a disastrous economic period, along with a period of political turmoil culminating in the overthrow of the Marcos regime. There were many pressures on the new Aquino government to act quickly. In 1988, the Comprehensive Agrarian Reform Program was introduced, intended to make 120,000 hectares available to the people of Negros Occidental on 30-year purchase terms. Similarly, an Integrated Social Forestry Program was introduced to provide a better base for managing the uplands, growing largely out of widespread ecological concerns. It made land available on 50-year leases in such a way that people would have a livelihood while the ecology of the uplands would be improved. Other programs were concerned with expanding irrigation, encouraging micro-enterprise development, and expanding the outreach of social services. All of these schemes were national. Negros Occidental differed from most other areas only in the way in which it incorporated the use of such schemes within an overall development strategy of an innovative nature.

Although the plans and programs looked fine on paper, it was clear that their implementation would require a mechanism. This mechanism had to reflect political commitment to the reforms envisioned, had to facilitate cooperation between the many parties that would need to be involved, had to make provision for the fact that survival dominated the lives of most rural families, and had to enthuse, motivate, or cajole many comparatively reluctant partners to participate in a process of change. In Negros Occidental, that mechanism was what became known as the Food-for-Work Program, although its scope would extend well beyond the provision of food in exchange for work.

The Food-for-Work Program was a large-scale and very comprehensive program of rural development, focusing on rural poverty alleviation. It came under, and was located within, the Provincial Governor's Office, which, along with a highly supportive governor, was one of the reasons for its success. Running for five years from 1990 to 1995, the program was funded by major outlays by the government of the Philippines, the World Food Programme, the United Nations Development Programme, and, to some degree, the Provincial Governor's Office. The program was to reflect and incorporate the national development schemes mentioned above — agriculture, social forestry, infrastructure, irrigation, and micro-enterprise development. Targets in each of these areas, in terms of "man-days," were set, that is, days of labor contributed by local people, especially the most needy, and funded by a rice allocation of two kilos per day or a combination of rice and a cash payment.

As we shall see, the food-for-work element was to prove crucial. Even though the projects undertaken had the potential ultimately to contribute to income generation, they could not do so in the short term, and poor people could not forego the task of feeding their families in order to construct an irrigation scheme, plant thousands of seedlings, attend an education program, or build a school for their children.

The food-for-work element was important; however, if this program had been restricted to the use of rice and cash to pay for the services of the poor, it is debatable that Negros Occidental would have started to change as dramatically as it did. The secret to the success of the program lies in the details of its implementation, and it is on those details that this chapter will focus.

THE OVERALL PROGRAM

This program is what I have described elsewhere as a comprehensive poverty alleviation program (United Nations ESCAP, 1996). It is comprehensive because it involves, in a collaborative manner, all of the necessary players and each of the relevant levels of change for comprehensive development to occur. The players range from the United Nations and national government, through the offices of local government (LGO) and government agencies (GAs), to the nongovernmental agencies (NGOs) and people's organizations (POs), with each category having a crucial role to play. The relevant levels of change include the central political and economic levels, the intermediate GA and NGO levels, and the local or grass-roots level (POs). The process is to some extent a combined top-down and bottom-up approach. It is motivated and facilitated by a central pivotal body, the Program Team, while also being a cumulative approach with successive dimensions added to what is essentially a local-level, community-based or people-centered development process.

This comprehensive nature of the program is crucial, but comprehensiveness does not emerge solely from a combining of particular components; it is dependent almost entirely on the way in which these particular components are molded together, and we shall consider the organizational aspects of that in the next section and the strategic aspects in the following ones.

A further crucial element of the program as a whole was that it was resource intensive. Large sums of money, large amounts of non-monetary resources, and significant inputs of knowledge were all significant in the success of this program. Being resource intensive was significant in such terms as minimizing corruption and avoiding a sense of participant dependency on aid. Indeed, it can appear inherently contradictory to use enormous resource inputs to inculcate participatory and self-reliant development at the local level; the answer to the implied dilemma, however, lies in the manner in which the resources were utilized.

On the other hand, if this type of development requires massive inputs of various types of resources, is it a feasible form of development for a range of settings? Are these resources likely to be available if the program were to be replicated in other places? The answer is no if this program as such is the model. It may be, however, that we shall be able to see ways in which the essentials of the model can be utilized on the basis of considerably reduced resource inputs.

THE ORGANIZATIONAL ELEMENTS

The program operated basically at three levels, each of which was crucial to the integrity and success of the program.

At the core of the organization was the Project Team, employed by the province and located in the governor's office. The team was made up of 18 young and enthusiastic university graduates, mainly from outside Negros, whose expertise spanned a range of disciplines. They operated as a team under charismatic leadership. They each knew what the others were doing and engaged in mutual support; they all shared their expertise at all levels, so that the team members' output was far more than the sum total of their expertise; and they presented as a coherent, enthusiastic, dedicated, and competent whole, inspiring all parties engaged in the program. Their style of operating was to be highly mobile, very accessible, and very concerned about outcomes. Because such a style renders the individuals very vulnerable to burnout, should they be without the support of a team, the team structure was vital.

Of fundamental importance at this level was the team's conscious reliance on others to introduce change. Team members were the enablers, the facilitators, and the motivators; they could not afford to allow themselves to be implementers. In any case, their ongoing presence was unlikely as the program was a five-year effort, and the sustainability of their work depended on the extent to which they could involve others in the actual development. Hence the team's essential task was to build a network of Local Implementing Agencies (LIAs).

The second organizational level was thus the network of LIAs. Some of these were NGOs, but the majority were GAs. Each of these LIAs agreed to play a role in the implementation of the program, but often, especially for the GAs, this role represented a radical departure from the way in which they had functioned in the past. Hence most of them required considerable input from the Program Team, an input designed to motivate, train, and support the LIAs so that they would realize their potential to contribute to comprehensive development and undertake the changes necessary to do so.

Not only, however, did these LIAs have to learn to adopt different strategies in their work; they also had to appreciate the need for, and learn the skills of, collaboration with all the other LIAs engaged in any specific community. In many situations, and the Philippines was no exception, GAs are more likely to compete with each other for power and resources than to collaborate, whereas NGOs and GAs are inclined to build a barrier of mutual distrust between themselves. If development was to make real inroads into Negros Occidental, these rivalries and barriers had to be removed, or at least minimized, for it appeared to be difficult to change entrenched attitudes over a short period.

This organizational level was built, strengthened, and maintained initially by the Program Team. It arranged a carefully planned series of monthly meetings designed to promote collaboration, although the actual collaboration of the LIAs in the field eventually became the main vehicle for the building of bridges between the LIAs. This collaboration was not only central to the program but was an essential ingredient in the future development of the province.

The role of the LIAs was, as their title implies, to be implementers. Each LIA was therefore responsible for a series of projects, endorsed by the Program Team (especially if a food-for-work element was involved) and implemented with the team's support and invariably the collaboration of other LIAs.

The third crucial organizational level was the community level, usually operationalized in the form of POs. Sometimes POs were in effect the LIAs, but most frequently they represented the element of people's participation which lay at the heart of the basic approach to development adopted in this program. Sometimes the people would suggest the project, with one or more LIAs and the Program Team responding to their initiative; sometimes a potential LIA would propose a project, but it would then proceed only with the people's endorsement; and sometimes a community development initiative was called for if a community and PO were to emerge to participate in their own development. Some of the NGO LIAs were critical of the inability of GAs and others to appreciate this last approach — the need for slow and painstaking work to build a population's capacity to function as a community through its own POs.

The organizational component of the program revolved around outreach, awareness raising, and collaboration. Outreach was the first requirement. The Program Team needed to reach out to and recruit a network of LIAs, and it also needed in the course of the implementation of projects to reach out to the participating POs. The LIAs, in turn, had to reach out first to each other, to the extent that most projects required such collaboration, and second to the POs that were their partners in the field. Finally, the POs needed to acquire the skills and confidence to reach out to the network of GAs, NGOs and LGOs, all with the capacity to support their development.

Outreach, however, had to be complemented by awareness raising. The Program Team, as young graduates, needed to inculcate an awareness of the realities of GAs, NGOs, and POs; the GAs had to become more aware of the objectives and constraints of each other; and POs often had much to learn about the variety of organizations potentially able to assist their development.

A crucial aspect of awareness raising was awareness of the nature of the development enterprise. It had been all too easy for the GAs, for example, to perceive their roles in highly concrete terms. They had to construct schools, or build roads, or introduce irrigation, and so on. There was often little awareness that the process of carrying out such roles was itself important; that the end products, schools and roads for example, were far more than physical structures in terms of their role in development; and that only when GAs' tasks were integrated with those of others would changes begin to emerge. For example, awareness lay in appreciating the importance of the local people's involvement in every facet of a school's construction; awareness lay in appreciating exactly how a road would contribute to a community's development, with the details of the project being consistent with maximizing development potential; and awareness lay in realizing that an irrigation scheme, to be successfully used and maintained, had to relate to the people's appreciation of its contribution to income generation, to the roles of other GAs in assisting agriculture, and to the people's ability to make the scheme sustainable into the future. This awareness raising constituted a major and ongoing role of the Program Team and, increasingly, of key personnel in the various LIAs. It required, however, an organizational structure to achieve that end. The nature and role of the team was pivotal, but the monthly meetings with face-to-face interaction and the skillful linking of the appropriate organizations into each and every project were the major organizational strategies. The team could use the meetings and the project development work to facilitate ever-growing awareness, until it generated its own momentum.

Only with outreach and awareness being achieved at certain levels was true collaboration possible. The program rested quite clearly on an appreciation of this fact, yet in practice creating an adequate basis of outreach and awareness was not always achieved, and collaboration was not then obtainable to the degree necessary. While difficult to achieve, the key ingredient of the program was collaboration. Development emerged from partnerships between GAs and NGOs, between POs and their communities, between people in communities so that strong and healthy POs would emerge, and, of course, always between the Program Team and all of the constituent parties.

Another level at which awareness was crucial was in relation to ecological issues. The importance of reafforestation of upland areas, of devising ecologically appropriate methods of managing the development of steeper upland slopes, of constructing ecologically sustainable irrigation systems, of establishing responsible logging practices, and so on all formed a central element in the awareness-raising program.

THE SPECIFIC ACTIVITIES OF THE PROGRAM

As pointed out early in this chapter, the program was to reflect national objectives in relation to agriculture, social forestry, infrastructure, irrigation, and micro-enterprise development, and the program was designed around set goals in each area. On paper, the objectives were to construct schools, irrigation schemes, stretches of road, and so on, utilizing a set number of "man-days" involving food and/or cash payments to achieve these objectives. Moreover, formally it was this set of objectives which was followed, and achievements could be and were officially presented in terms of number of schools built or number of trees planted, together with the food-for-work components presented in terms of "man-days." Both objectives were indeed important. Each of the respective concrete achievements could be seen as a direct and tangible contribution to development, and the "man-days" represented opportunities for reimbursed work for the poorer members of many communities around the province. No one outcome, however, had the potential in itself to alleviate poverty. All were certainly capable of contributing to that ultimate goal, but as such their contribution could be significant or minimal. It is necessary, therefore, to look beyond the achievements reported in the program reports and see the specific activities in their full and complex nature.

The program was, in terms of specific development outcomes as distinct from organizational building, composed of a large number of projects, reflecting the different dimensions of the above-presented objectives. In 1994 there were 63 project sites, with a much larger number of specific projects. The procedure for the initiation of each project was basically the same. The initial proposal might come from a PO, a GA, or an NGO. If the initiator was not a PO, the community's endorsement was vital. The project was then carefully assessed, with modifications often made, before it was approved as part of the program. The process was highly consultative, with the ultimate criterion for approval being the project's capacity to enhance the community's well-being. The ultimate aim in relation to each community was to contribute to meeting that community's needs. Invariably, such projects represented a proactive approach on the part of the team or an LIA, while others were a response to community initiatives.

Let us consider specific projects initially from the perspective of particular LIAs, selecting a few examples. The Office for Provincial Agriculture had 32 current projects at the time of the visit, almost all of which were income generation in nature. The initial scheme often involved the use of land acquired by government and made available to selected families. The farmers would receive training under the food-for-work arrangement, would be supported in planning the utilization of the land, and would be able to join in marketing and credit schemes. The office saw its role not as providing for beneficiaries but as empowering farmers to participate fully in all dimensions of agricultural development. It saw motivation and people's full participation as crucial to success. Organizationally, it had come to appreciate the importance of flexibility in its work and of devolving management to the field level. However, it was quick to acknowledge that the food-for-work element was essential to enable training and development, with longer term results to take place. Also, the integrated approach was fully appreciated in a way in which, by the office's own admission, it had not been in the past. The lament in the office was that other elements of the bureaucracy were not as ready as it was to embrace the new ways of operating.

The Education Department faced a situation of an inadequate infrastructure. Negros Occidental had 3% of the country's population but only 1% of its schools, and many communities were without a school. In the past, the department had identified the deprived communities and contracted construction firms, which used outside labor, to construct as many schools as its budget would permit according to a predetermined model. Only a relatively small number of schools were constructed in this way. They were often unfinished in terms of furnishings, and many were not utilized, due partly to the shortage of teachers and the lack of local motivation to organize substitute ways of utilizing them. Following the department's involvement in the food-for-work program, this situation had changed radically.

In 1994 the department had 59 project sites — a significant increase in the school construction rate. The projects operated through local parents' associations, which for the most part had requested the school and were managing the building of it. Use of the food-for-work provision meant that poorer local families carried out the building under supervision. By so doing, they were learning new skills, increasing their self-confidence, and developing a new image of themselves as community members. By contrast with past experience, the quality of the construction was high and the degree of finishing touches far superior. Moreover, the cost was less and corruption almost completely eliminated. Most importantly, department staff had learned to appreciate the importance of relating to other GAs, in relation to matters such as children's health, the use of plantings to enhance the environment, and the relationship of the education to be provided to the income-generation realities and socioeconomic needs of the community.

The Provincial Engineers Office had already paved 6,000 linear meters of road, and in doing so had been able to focus on the critical sections of roads. Using local labor under the food-for-work scheme, seen as critical to its achievement of high targets, it had become very conscious of the diverse roles of paved roadways. Of course roads were to facilitate movement of people and goods, but roads could be vital also to income generation, to agriculture, to the provision of social services, and to community building. For example, paving a section of road through a settlement provided many services: it made movement possible in the wet season, it provided a platform for drying grain, it gave the children a hard and flat surface for recreation, it helped to bind the community together, and so on.

The Provincial Water Taskforce was responsible for wells, irrigation, footbridges, and the like. In the past its role had been the provision of such to sites where the beneficiaries carried some influence. As a very active participant in the program, its whole approach had changed. It now selected its projects on a basis of need, worked closely with the local community, ensured that the local government would maintain the resource provided, engaged in considerable education and training around the projects, and was far more conscious than before of the ecological aspects of what it was doing. Task force members were now proud, in a way that they could seldom be in the past, that they had been instrumental in producing an asset which was pivotal to the quality of life of significant numbers of families.

Each of the above examples is of a GA operating as an LIA in the program. There could be no doubt that the very nature of the GAs in Negros Occidental, in terms of perception of roles, internal organization, and methods of operating, had changed dramatically. Moreover, the changes had grown out of enhanced awareness and seemed in large part to have evolved over time. Some workers suggested that a three-year development process had been required to achieve such dramatic change. Most importantly, however, was the obviously high level of work satisfaction emanating from the new ways of working. Awareness, motivation, enthusiasm — these were the best descriptions of what one met in these offices, and it boded well for the future. But it was no accident. The civil servants were unanimous and spontaneous in giving credit to the team for having achieved this outcome. Call them catalysts, motivators, facilitators, or what you will, the qualities of the team members, together with their skill at outreach, awareness raising, and enhancing collaboration, were crucial to the success achieved.

Much fewer in number, and in their eyes somewhat devalued in the overall program, were the NGOs which functioned as LIAs. Although the NGOs were varied, and played a variety of roles, their perception of the task of poverty alleviation tended to be essentially different from that of the GAs, as were the levels at which they were able to contribute to specific projects.

One role for NGOs as LIAs was community development, or enhancing the capacity of what might initially be a collection of disparate families to form POs and engage in project work. Such NGOs tended to see the GAs as underestimating the importance of facilitating PO formation and often the painstaking work required to achieve this. One NGO spoke of a three-stage approach: (1) immersion in a situation to establish rapport; (2) participation with the community in appraising situations, planning activities, and engaging in projects; and (3) phasing out — allowing the community as it had now become to operate in a self-sustaining and community-building process. This whole process could well require a worker to virtually reside in the project location for a year or more. In this type of work, the food-for-work program had a negligible role to play; it would come into its own as the community became cohesive and began to initiate projects. In this way, food-for-work was a supplement, not an alternative, to development. These NGOs were also highly conscious of the importance of

land reform, of an ecologically sound approach to development, of the importance of GAs' roles, and of the significance of the overall program as such. As
they saw it, awareness raising and local organizational development were the
essential foundations of development, but without the provision of resources and
real opportunities for income generation and so on, these foundations were
likely to generate frustration and anger rather than positive and constructive
progress.

One NGO had, under the program, worked with 19,900 beneficiaries over
some three years, with a variety of schemes and areas of operation. Of its 19,900
beneficiaries, some 1,000 had participated in food-for-work programs. While
significant for the individuals and the projects, these figures reflected the fact
that the work was far broader than the name Food-for-Work Program implied.
This NGO, whatever the final nature of the project, focused on local organizational development and general capacity building, although its key objective was
always livelihood support, as it termed it, or income generation. Training was
always strongly emphasized and invariably covered a range of general skills as
well as being project-specific.

Among the projects in which this NGO participated, along with the people,
were improving fishing through boat and artificial reef construction, promoting
integrated farming on an organic basis, and a self-help housing scheme utilizing
mass-produced building materials and a carefully developed design. But whatever the project, the fundamental need was seen to be for community development and PO building, including helping to weld POs into networks.

We have considered some specific activities from the viewpoint of the LIAs.
Their role was, of course, crucial, and their perspective is extremely important
because, to a significant degree, the sustainability of the changes which were
occurring across the province depended on fundamental changes in the network
of GAs and NGOs, through which most of the available resources would continue to flow. However, in the philosophy of complementary top-down and
bottom-up development, the role of the POs would also be crucial, so let us
consider a specific project from the community's perspective.

This project started with the identification of nine households engaged in
illegal logging in an uplands area and having virtually no access to social
services. In 1991, through the outreach of the program, they were offered
uplands land on the basis of a 25-year lease with the right to renew for a further
25 years. The households accepted the offer, and the task of building disparate
households into a unified and viable community commenced. The number grew
quickly to 25 households, and these households began the work of community
building. In this they were aided by the deployment of an NGO staff member
who worked with the people as a community development worker for well over
a year. All parties regarded the role of that person as pivotal, for these households had a limited capacity to envisage the potential of the land and the
importance of community building.

With the assistance of the catalyst, the households gradually evolved a plan for the establishment of their village. It consisted of a series of specific projects. They would construct a dam and a mini-irrigation scheme to bring water to rice paddies. With this irrigation they would develop an area of intensive cultivation, producing rice, vegetables, and flowers which could be sold as cut flowers to the people on the coastal plain. A further area would be devoted to forestry, partly to yield a longer term income and partly to restore the area's ecological balance. Much of this work, however, would require skills which the people did not possess, so training and education were important. Finally, the households needed to become a community, served by a school and possessing a range of other facilities in the area where their houses were being erected.

All in all, it was a mammoth undertaking for these people. Even trusting each other as the basis for engaging in joint community exercises was difficult, and some friction, distrust, rival leadership, and so on were not uncommon during the next three years. However, the households did gradually learn to work together, to establish their community structures, and to engage in collaborative measures. This was greatly aided by the resident catalyst, but it was also the experience of working together on joint projects which built the community. These joint projects represented the pooling of labor on community schemes, facilitated by the food-for-work arrangement; the regular site visits of team members; the support of a range of GAs, acting as LIAs on specific projects; and the whole experience of learning together, both in classrooms and within the project work.

In three years, a disparate collection of impoverished and marginalized households had become a flourishing community — flourishing socially and economically. Their dam was both beautiful and the heart of a very successful irrigation system. Their fields were highly productive, and thousands of small trees dotted the reafforested area. Their children participated in school, and the village was a happy and well-serviced place to live. Almost daily a community person would head off with the cut flowers to sell, while others pursued other income-generation activities, some operating on an individualistic basis and some on a cooperative basis. To the visitor from outside, there was no mistaking the people's pride in their achievements and their ability to function successfully as an economic and social unit.

Each project had involved them in learning skills, formally and informally; each one had helped them to understand the work of one or more GA, or other sources of resources, and to gain the knowledge and confidence to access them; and each one had drawn them closer together as a community. All of these experiences constituted the basis for the future. The catalyst was crucial but could then withdraw; the food-for-work arrangement was indispensable for the initial labor-intensive tasks, like the construction of the irrigation scheme, undertaken by families dependent on their daily activities to survive, but it soon became dispensable. The team visits were very important in the early period for welding the projects together and supporting the community, but by 1994 the team members were friends rather than facilitators and able to concentrate their time elsewhere. The network of GAs and NGOs was now an understood source of resources, involving not charity but services to be purchased, argued for, or manipulated by the people as citizens of a province and a country.

It was a long way from this people's isolated, marginalized, and disadvantaged status of the past. It had taken but three short years to achieve the change, but the distance traveled and the investments from all concerned in reaching the end (or was it the beginning?) were enormous.

PRINCIPLES UNDERLYING THE PROGRAM

To the best of my knowledge, no one connected with this program of poverty alleviation in Negros Occidental had enunciated a set of principles on which the program would be based, or indeed a set of strategies by which it would be pursued. The program was essentially a food-for-work program designed to make possible the implementation of a range of government policy initiatives, thereby alleviating the rural poverty which characterized the province. The team established to implement the program would, however, seem to have devised and pursued a set of directions which, to this group of intelligent and educated young activists, appeared to be the logical way to proceed. No doubt program consultants and others around them contributed to the process, but essentially the program was based not on prior learning, experience, or prejudgments, but on an assessment of the presenting realities. That this was the case and that the program was so clearly successful renders highly significant the principles adopted and the strategies which emerged.

At the very heart of the program was a set of principles which signified a belief that successful development and poverty alleviation had to be people centered or community based. The three key principles were participation, self-reliance, and sustainability, culminating in a focus on people in a community.

To be approved under the program, every project had to have the full support and potential full participation of the people in a particular location. In a very real sense, it was their project, to be implemented with the assistance of an LIA. Every project's implementation had to be carried out through and with the people, organized within a PO. Especially where the food-for-work arrangement was being utilized, the labor for projects was to be provided by the poorer members of the local community, offering them a temporary respite from poverty together with the chance to acquire some skills and some sense of status and purpose (self-esteem building). However, people's participation went well beyond this level as POs were invariably partners in all project implementation. Moreover, the whole approach or process was of a participatory nature, so that people were not only participants at an organizational level but clearly were, and felt themselves to be, full partners.

Self-reliance was the second crucial principle, reflecting the focus on people in communities. Every step of the program was designed to maximize self-reliance. Where NGO staff worked as community developers, it was geared to their eventual withdrawal. The relationship between POs and local GAs was never dependent in nature, but designed to inculcate in POs an ability and a willingness to negotiate with any GA for access to its resources and services. Training was designed to give people complete control over the technology they were utilizing, as well as the knowledge essential to their development. The team was to be there for only an initial period, ensuring that developments at any level did not depend on them. Self-reliance was fundamental to the program at every level. Self-reliance is also an important principle in relation to ecologically sustainable development, in that it makes people more aware of the need to nurture carefully their local natural resources as crucial to their livelihood.

The final principle in this basic trio was indeed sustainability. The program rested on changes at the local community level, changes in the GAs and NGOs which served as LIAs and changes in the province's overall arrangements in relation, for example, to land ownership and utilization. At each of these levels, the emphasis was on ensuring that the changes were sustainable. This was achieved in part through embracing a participatory approach and in part by ensuring that at all levels the knowledge and understanding of what was being implemented were there. Within each of the projects, sustainability lay in ownership of the outcome, be it a parents' association, a cooperative, or a PO; in a full understanding of the end product; and in arrangements for maintenance of the project.

Behind each of these three principles, the focus on the people in poverty and their organization into communities and POs was pivotal. Essentially participation, self-reliance, and sustainability rested on the proposition that poverty alleviation had to be based on the empowerment of those in poverty and that poverty alleviation, despite the many and various inputs required, was ultimately a process to be undertaken by the people themselves. Development of the people into communities and organizations, where necessary, would often require the services of catalysts, facilitators, and a network of collaborators, but ultimately, if the people did not elect to initiate change of their own free will, little would change.

STRATEGIES EMPLOYED IN THE PROGRAM

As with the principles, the strategies which were clearly obvious to an observer of this program, and affirmed in discussions, had evolved with the program. There was no blueprint of strategies preexisting the program. This fact renders these strategies all the more important, for they were strategies found from experience to be crucial in the field.

The core strategy was the use of a team as the program catalyst. This was a situation where the catalyst had an important function. For example, the team found that the concepts of and approaches to development and poverty alleviation which they adopted required some three years of hard and continuous effort for the GAs to understand. Concepts like working through the people rather than providing something for them were new, approaches like collaboration between GAs themselves and GAs and NGOs had been alien to participants, and focusing on the poorer people as the major source of labor was also a new concept.

The team worked as a catalyst to stimulate participation at all levels in this extremely ambitious program. The team required certain characteristics to undertake and sustain such work. The fact that the 18 team members were young graduates was important, as was their diversity in terms of background disciplines. Many people in the LIAs and in the field commented on the team members' commitment and practical problem-solving ability. While that could be applied to the team as individuals, it was apparent that the team as such incorporated the knowledge of the disciplines represented and proceeded to build on that base a wealth of practical or applied knowledge, developed in the field and shared through regular discussion. In the end, team members said that they did not work primarily as representatives of their particular discipline, but as development workers — as team members engaged in a common development program. They had absorbed knowledge from each other and from the work itself, so that many described their work as 50% or more social work or development work and the balance agriculture or engineering and so on.

It was clearly crucial to the effectiveness of the program that the team function as a team, and not as a collective of individuals allocated specific roles. The team met with LIAs to plan projects and discuss any problems arising, visited project sites and discussed what was taking place with the POs, and sought to be sensitive to and supportive of the needs of individuals at all levels. Although in radio contact with the team's headquarters while in the field, team members frequently operated alone, in sometimes stressful situations, and were regularly confronted with situations calling for a wide range of skills, knowledge, and initiatives. Flexibility, versatility, and inventiveness became important assets in the field.

These common roles of team members, and the team members' ability to manage them over time, required that the team become a team. It needed to strengthen its members, engender mutual support, constitute a learning environment, and enable members to relax and switch off. The team's routine therefore included regular sharing of field experiences, training sessions run by outside consultants, periods of socializing, and opportunities for individuals to interact with and support each other. In physical terms, the layout of headquarters and the choice of places for retreat were important. Above all perhaps, leadership was important. The team leader needed, on the one hand, to be part of the team, sharing the team members' basic characteristics; on the other hand, he (a man in this case) needed to be charismatic, instilling in members a sense of security, stability, support, and concern.

At the wider level, the team required an organizational context which would enable it to operate as a team. It was, after all, a creation of government, and it could have been set up as a group of public servants under an assistant director within a larger government department — the situation of development workers in some countries. Based on someone's insight, or fortuitously, this had not occurred here. The team was located in the Office of the Provincial Governor, with the emphasis on providing support to a relatively autonomous, but of course accountable, entity rather than direction.

A significant set of strategies could be seen as constituting a development model. It seems unlikely that the model effectively adopted here was a textbook or experience-based one, simply adopted and applied to the program. The model evolved as the common sense path to development as perceived by a group of intelligent workers in consultation with many others. Let us examine this set of strategies.

At its heart was participation as a strategy, so that development was people centered or community based. Second, the people needed to function on an organized basis, so the establishment of POs was essential. While POs had to be the initiative of the people, their emergence could be facilitated or encouraged by catalysts, engaged in a strategy referred to variously as community development, social mobilization, or awareness raising — a role perhaps best played by NGO staff.

Once a community was ready to embark on its own development, and sometimes even before, the strategy of education or training was seen as essential. It was not sufficient that people function under direction. They needed to be fully involved in the planning and implementation of all formal projects, and to do so required extensive education or training. The training provided was of various kinds, but all were important. Formal training was usually part of a food-for-work project. A group of people would be offered training in a particular field, with the food entitlement enabling them to continue providing for their families. Informal training occurred whenever a team member or LIA staff member visited a project site to discuss progress and any difficulties encountered. Non-formal education occurred when the people sat down, with or without an outsider present, to work out a plan of action. There was often within the group considerable knowledge to be shared and always the potential to develop further knowledge.

A fifth strategy, which emerged spontaneously from social mobilization and training, was income generation in its various aspects. Central aspects of awareness raising, of PO formation, and of training were in fact related to income generation, which was not surprising given the poverty alleviation objective. What was important, however, was the component parts of the income-generation strategy, and they varied somewhat from project to project. Among them, however, were the ability to (a) identify a potential income source, (b) assess its viability from a marketing and financial perspective, (c) develop sources of credit and the ability to use them soundly, (d) develop a plan which included longer term income sources (e.g., forestry) and the essential sources for meeting short-term needs, and (e) appreciate the techniques of and develop the confidence to reach out to any source of resources available to them.

THE FUNDING BASE AND ITS IMPLICATIONS

It is clear that a program of this kind calls for a considerable input of resources, financial and otherwise. The team of 18 with its transport, radio system for maintaining contact while in the field, office base, and so on was one major expense, although when this sum is divided by the large number of projects successfully initiated, that cost seems small. The Food-for-Work Program was a second large expense, intended to facilitate 6.5 million "man-days" of labor and training. The benefits of this program were recognized by all connected with it. It was an ideal use of the World Food Programme's resources. Other expenses were related to the many other aspects of the project work undertaken by the LIAs.

The dilemma arises when the number of situations worldwide able to benefit from the type of program outlined is multiplied by the cost involved. Presumably, with careful use of a combination of local, national, and international resources, many such programs could be run. Moreover, not all situations, unfortunately, will be conducive to the type of program discussed here. A significant question which is still important to consider, however, is whether a program such as this could be introduced on a less resource-intensive basis.

The size of the team is relative, of course, to the number of projects to be introduced. However, the qualities of team members and their ability to function as a team should not be jeopardized. Salary level may not be a major concern to some young graduates, but if their contribution is desired for four or five years, it cannot be too low. The use of a food-for-work operation will depend on the extent of poverty in an area. What is absolutely clear from this program is that where poverty is extreme, the ability of the poor to participate in work or communal projects or in training is limited unless it is facilitated by such as a food-for-work program. Support for their families' survival is crucial to their participation, and their participation is crucial to the projects.

Thus, the two major costs of the program would seem crucial to success. Costs associated with the individual projects, apart from the food-for-work element, should be borne by the GAs initially and in time by the POs, in the spirit of self-reliance and sustainability. It is a mistake to imagine that development can be carried out effectively on the cheap.

CONCLUSIONS

A major challenge confronting the contemporary world is poverty alleviation, and successful poverty alleviation can ultimately be achieved only by people working in communities, but working in partnership with others. Community development, in the sense of the forming of communities and their organization into POs, is the foundation of successful poverty alleviation. Its full achievement in most poverty-stricken situations will require the full participation of a range of NGOs and GAs, working in such close collaboration with each other that the ensuing development initiatives are well integrated within an ecologically sustainable framework. The whole process, however, of involving a range of agencies in a people-centered or community-based development program, which is well integrated, requires the work of a catalyst. In this program, that catalyst was a team of young graduates. Their role was crucial in motivating the various agencies' involvement, in inspiring commitment from a variety of individuals, in helping all parties to appreciate the nature and importance of people-centered development, and in maintaining the overall direction of the program.

It is apparent from a study of this and other poverty alleviation programs that rural poverty alleviation (and perhaps also urban) is most successful when based on certain principles and implemented according to certain strategies, as are outlined in this chapter. To be fully successful, however, the program should occur within a context of political stability and be well resourced. The Negros program enjoyed both of these conditions. The absence of one or both of these conditions does not render impossible community-based poverty alleviation. It may make the work more difficult, and it may place limitations on the outcomes to be expected, but it will also render even more important the community focus of this work. Ideally, development is a combination of top-down and bottom-up initiatives. If the extent of the top-down contribution is limited by political instability or the availability of external resources, the role of the community is that much more important.

The key to poverty alleviation is ultimately sustainable community development. If that development is buttressed by a network of government service agencies committed to sustainable community development, and by a political system geared to providing an enabling environment for development, the potential of community development is greatly enhanced. Sustainable community development itself, however, must ultimately be manifested in effective POs, initially established for development purposes but ultimately finding their place within civil society, which is that level of community development which links the needs and wishes of the people to the macro structures and policies of society.

AUTHOR'S NOTE

The author carried out an on-site assessment of the program described in this chapter in 1994. The assessment formed part of a study organized by the Social Development Division of the Economic and Social Commission for Asia and the Pacific (ESCAP, 1997) and coordinated by the author. The study, across five countries, was designed to identify the strategies which contributed to successful rural poverty alleviation for training purposes. The author was assisted in carrying out this work by Mr. Sarathchandra Gamlath of Rahuna University in Sri Lanka.

REFERENCE

United Stations ESCAP (1996). *Showing the way. Methodologies for successful poverty alleviation projects, No. 1, Promoting human resource development services for the poor.* New York: United Nations.

CHAPTER 4

The Yawanawá–Aveda Bixa Project: A Business Partnership Seeking Sustainability in an Amazonian Indigenous Community

Sandra De Carlo and José Drummond

> One indigenous community alone cannot contribute significantly to the global process of environmental solutions for the planet. But I think that the whole of indigenous thought, the memory and traditions that our people have of nature and human life are the most important foundation for orientation towards sustainable development and the adequate management of natural resources....However, if this knowledge is fragmented, then it does not make sense. Do you understand the complexity of this? As a whole it makes sense, but fragmented it does not, and is self-destructive.
>
> —Brazilian Indigenous Leader[1]

The Brazilian Amazon region has suffered a disorderly process of development that has affected not only its natural ecology but also the cultural and social dimensions of its inhabitants (indigenous peoples, *caboclos,*[2] urban dwellers). Therefore, when considering any initiative to be included in sustainable development for the Amazon, one must look at the region as much more than a natural sanctuary to be preserved in its entirety without respect to the demands of its local population (Kitamura, 1994). This chapter focuses on a case study of a company working with an indigenous community on a sustainability initiative in the state of Acre, Brazil.

Historically, communities engaged in trade economies have depended on outside markets. Today's trend toward the concept of sustainable development for the Amazon has the same rationale, that is, creating wider markets for the region's natural resources. The only difference is that currently this discourse has an ecological approach. In this context, "green" and socially responsible businesses are marketing foods, cosmetics, soaps, buttons, and other products from the tropical rain forests. The underlying idea is that these ecosystems can be economically productive by involving communities in ecologically sustainable types of production. Forest-based enterprises are gaining acceptance among local communities that seek economic alternatives and among international investors that hope to improve their environmental records.

However, it is important to establish clearly if these projects are moving toward sustainability for the communities involved. It is necessary to record the experiences of communities that seek to engage in diversified production directed to wider markets without disrupting their cultural values. This study uses field observations, interviews, and documentary sources to record the experience of a specific indigenous community, the Yawanawá, in building a business partnership with an American company (Aveda Corporation). The Yawanawá–Aveda Bixa Project exemplifies many issues in the debate surrounding the concept of sustainability for rain forest–based enterprises in remote communities that seek "green" markets.

GEOGRAPHIC SETTING

The Gregório River Indigenous Reserve covers 92,859 hectares and is located at the headwaters of the Gregório River, in the west corner of the state of Acre, approximately 500 kilometers from Rio Branco, the state capital (see Figure 4.1). Acre is the westernmost Brazilian state, with an area of 153,736 square kilometers, occupying 3.2% of Brazil's legally defined Amazon. There are two watersheds in Acre, corresponding to the Alto Juruá and Alto Purús rivers, both major tributaries of the Solimões and the Amazon rivers. The state's natural vegetation is primarily evergreen, tropical rain forest characterized by heterogeneity and diversity of plant species. Only 6% of the state's forest cover has been cut down (SEPLAN, 1993). The Gregório River, a tributary of the Juruá, is a small to medium serpentine river with red/brownish waters. Its water volume varies considerably between the rainy (October to April) and dry (May to September) seasons. During the months of August and September, the river is rarely navigable due to low water levels. From October to May, higher water levels allow the traffic of larger boats. Annual rainfall is between 1,750 to 2,750 millimeters.

Two tribal groups with distinct cultural patterns share the reserve. Close to 600 Indians live there, about 400 Yawanawá and 200 Katukina. The Yawanawá population has been growing rapidly in recent years. Relations between the two groups are good, despite their differences, the most obvious one being that the Yawanawá have been more open to contact with other communities and ethnic

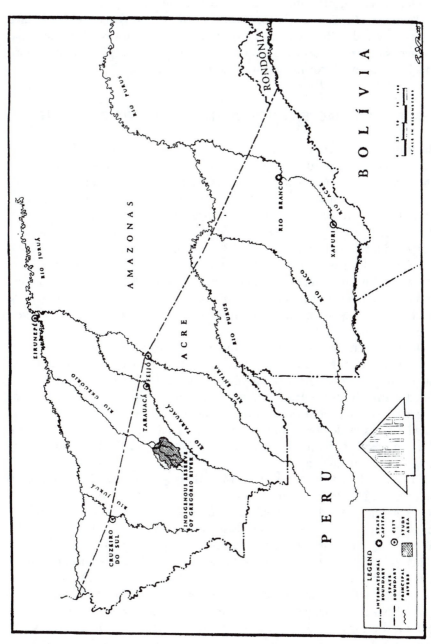

Figure 4.1 State of Acre, Brazil.

groups. The Katukina tribe hosts the New Tribes Mission, a Protestant group. It also has a landing strip in its village of Sete Estrelas.

The main Yawanawá village, Nova Esperança, is very remote. It can take three to seven days to reach the village from the major towns of Tarauacá and Cruzeiro do Sul, in the state of Acre, and Eirunepé, in the state of Amazonas. This is a limiting factor in the tribe's access to markets.

GENERAL CHARACTERISTICS OF THE YAWANAWÁ COMMUNITY

The Yawanawá are members of the larger Pano indigenous linguistic group and share the cultural traits of the Nawa nations (Kashinawá, Poyanawá, and others) who have historically occupied the forests and river valleys of southwestern Amazonia. They all lived in small tribal groups of about 200 to 250 people. These tribes were first contacted by Europeans or neo-Europeans in the late 1800s (Comissão Pró-Índio do Acre, 1996). The Yawanawá have been in regular contact with European culture since 1880, as a consequence of the rubber boom. This led to changes in their culture, with the introduction of new habits and dependency on goods such as medicine, ammunition, clothes, cooking pans, and so on. The houses of the Yawanawá villages are built on stilts and are crafted mainly from wood frame poles and from leaves of a coconut species (*cocão*), vines, and other regional plant species.

Language and religious rituals have remained largely the same since contact with Europeans. The Yawanawá speak their own language, as well as Portuguese. About 10% of the Yawanawá are Protestant. Their main ritual party is called *mariri*. In this ritual they dance, sing, and drink *cipó*, a beverage made from a type of vine. For this dance they paint themselves with *urucum*[3] and dress up with skirts made of a palm tree fiber *(buriti)* and hats made of *taboca* fiber. The remote village has no electrical power, although a small solar energy facility feeds their shortwave radio.

In terms of land use, the Yawanawá world is well structured. Social organization is integrated with adaptation to the environment. Even with rapid changes, they have maintained traditional subsistence activities. The Yawanawá grow several varieties of yams, manioc (cassava), sugarcane, rice, beans, and corn (Bona, 1995). They also grow pineapples and fruit trees such as bananas, avocados, papayas, mangos, and citrus. Some of the wild fruits gathered from the forest are (in Portuguese): *açaí, bacaba, jenipapo,* and *jaracatiá.* Since these subsistence activities are cyclical and seasonal, abundance and scarcity alternate. They keep cattle, pigs, chickens, and ducks. Hunting and fishing are other traditional means for subsistence. Favorite game animals include peccary (the tribe's name, Yawanawá, means "many wild-pig people"), tapir, deer, paca and *cotia* (rodents), *guariba* (a primate), turtles, and armadillos. As contact with non-Indians led the Yawanawá to hunt with rifles and made them dependent on ammunition, hunting now requires external commodities.

HISTORY AND OCCUPATION OF LAND

Slavery, destruction, and social disruption of indigenous peoples continued in Brazil long after the formal abolition of slavery in 1888. Both intertribal wars and diseases decimated native populations. With the rubber boom, rubber-tappers carried devastating diseases into the furthermost places of the Amazon basin. However, unlike so many other tribes of the state of Acre, the Yawanawá were not actively exterminated. For about 80 years, they traded rubber in the form of goods at the *barracões* (trading posts), where rubber-tappers bought their supplies on credit at grossly inflated prices. Until the mid-1970s, the Yawanawá worked as de facto slave labor for isolated rubber-tapping operations in the Caxinawá and Sete Estrelas areas within the current reserve. They performed many tasks for their employers, such as tapping trees, opening and maintaining trails, clearing brush, building houses and corrals, transporting rubber, and cutting timber (Comissão Pró-Índio do Acre, 1996).

In 1974 the Paranaense Company for Cattle and Industrial Colonization (PARANACRE), a ranching and logging company, bought dozens of rubber-producing holdings (a total of about 450,000 hectares) along the Gregório and several other nearby rivers. Local Indians were hired by the company as rubber-tappers, peons, and wage laborers. Fortunately, not much wood was extracted from the area due to transportation problems (Yawanawá, 1996). When the Gregório River Indigenous Reserve was first designated in 1977, the Yawanawá, together with the Katukina, organized to exclude PARANACRE administrators and other non-indigenous inhabitants from the area. In 1983, with the support of the Brazilian Indian Agency (FUNAI), the Missionary Indigenous Council (CIMI), and the Pro-Indian Commission of Acre (CPI-Acre), the Yawanawá and Katukina occupied the rubber holdings, recovered the collected rubber, and negotiated the removal of PARANACRE management, putting an end to the company's commercial monopoly. The creation of cooperatives managed by Indian leaders allowed more autonomy in the sale of extractive and agricultural goods and gave better access to less expensive and more diversified goods in nearby towns. In 1984, the Gregório River Indigenous Reserve was officially demarcated within the county of Tarauacá. However, PARANACRE still owns extensive properties around the reserve, leading to constant transit of outsiders through the reserve. As the first demarcated indigenous homeland in Acre, this reserve is a model for local Indian leaders in their efforts to claim title to their lands (Comissão Pró-Índio do Acre, 1996).

As part of these historical changes, younger Yawanawá leaders, along with those of other indigenous groups, began to attend courses given by CPI-Acre. With the creation of "forest schools" in 1984, indigenous instructors started to teach literacy to young Indians, reclaiming their traditions, generating bilingual texts, and preparing the people for more advantageous interactions with the non-indigenous society. These schools are a source of pride and reassertion of ethnic identity for the Indians.

Beginning in 1988, Yawanawá health workers were taught (with technical support from CPI-Acre) how to provide immunizations and basic healthcare, lowering the incidence of diseases and common epidemics that affect many indigenous populations. The Yawanawá cannot address these problems without outside help, and governmental agencies have neglected their medical needs. The Yawanawá suffer from both newly introduced diseases and previously existing ones. They try their traditional medical techniques on some health problems, such as heart disease, snake bites, and *ferida brava* (leishmaniose) and other skin infections. Healthcare remains insufficient, though, in spite of these and other serious health problems. Young leaders created a new situation by requesting the New Tribes Mission to leave their village. The Yawanawá are still besieged by missionaries, but the leaders continue to resist outside interference.

With the exclusion of PARANACRE and the decline of rubber prices after 1988, the Yawanawá began to look for new economic alternatives. At the same time, the tribe engaged in a project to cut lumber and sell hardwoods from its territory, through FUNAI, hoping this would be profitable. However, logging proved detrimental to the tribe's traditional uses of forest resources and also negatively impacted social and cultural realms. Since logging did not even provide sufficient economic returns, many Yawanawá began to migrate to cities such as Feijó, Tarauacá, and Rio Branco, in search of jobs. Worst of all, logging attracted outsiders to Yawanawá lands. Combined with the interference from missionary groups, this affected the cohesion of the Yawanawá. Logging also caused great controversy and quarrels among the natives, FUNAI, and environmental organizations, which accused the Indians of degrading their own environment (Yawanawá, 1996).

PROCESSES LEADING TO REVITALIZATION

These obstacles stimulated Yawanawá leaders to seek new directions to achieve economic and social revival of their people. Due to the shortcomings of government agencies, the Yawanawá sought financial and technical support in the areas of education and health. The first step came in 1992 with the creation of the Organization of Yawanawá Extractive Farmers of the Gregório River (OAEYRG). Through the organization, the leadership has represented the community at national and international levels, obtaining resources that have helped to expand and diversify productive activities and to continue education and health programs. These goals were implemented through partnerships with indigenous entities, humanitarian organizations, environmentalists, and private companies concerned with indigenous and environmental issues.

Yawanawá leaders are actively seeking ways to connect their villages to the regional economy without allowing the erosion of traditional culture. Among the difficulties are (1) long distances to the closest markets; (2) low prices paid for the products they sell, including rubber and handicrafts; (3) high prices of consumer goods, such as medicine, school materials, clothes, tools, gasoline, and durable goods; and (4) difficulties in dealing with a restricted market, rather than with cooperatively inclined partners (Comissão Pró-Índio do Acre, 1996).

At the moment, OAEYRG is carrying out two major income-generating projects: the Yawanawá–Aveda Bixa Project, which involves the planting of annatto and the extraction of bixin, and the Vegetable Leather Project, which involves the production of a rubber-coated cloth that can replace animal hides, plastics, or synthetic rubber. This product is sold to *Couro Vegetal da Amazônia,* a company based in Rio de Janeiro, which uses it to make handbags, key rings, hats, etc. It would be important to analyze both OAEYRG initiatives. However, due to time constraints during field research, this study focused only on the Yawanawá–Aveda Bixa Project.

THE YAWANAWÁ–AVEDA BIXA PROJECT

Yawanawá leader Biraci Brasil, who participated actively in the indigenous peoples' movement since the 1970s, was invited by the Coordinating Body for the Indigenous Peoples' Organizations of the Amazon (COICA) to participate at the Earth Summit Conference in 1992 in Rio de Janeiro. Together with Dionísio Soares (then a worker at the Acre Planning Secretariat), they conceived and wrote down a project based on the agroforestry concept of taking advantage of degraded areas to grow perennial species of commercial value. They focused on annatto (to extract bixin, a bright orange-red dye),[4] in combination with fruit tree species.

A basic goal of the project was to build partnerships with governmental and non-governmental institutions in order to obtain the proper technical and financial assistance. The choice of annatto was based on its steadily growing market demand. It is also culturally and agronomically appropriate to the Gregório River area and has regional markets. During the Earth Summit, Biraci Brasil met Horst Rechelbacher, founder and president of Aveda Corporation, a cosmetics company based in Minneapolis, Minnesota, and negotiations began. Because Aveda produces cosmetics based on raw botanical materials, it was interested in buying bixin. In 1993 Aveda acquired the rights to buy and resell annatto produced by the Yawanawá indigenous community. Additionally, the Yawanawá image could be used for Aveda's marketing (Arnt, 1994). In the United States and elsewhere, an organic product that comes from an indigenous community, with potential for protecting rain forests, is highly valued by many consumers of cosmetics. Aveda uses bixin to manufacture the Uruku™ line of "totally natural lip color," which sells for US$13.00 each.

The contract established between Aveda and OAEYRG spelled out the goals of both parties. The Yawanawá wanted "to re-vitalize their economic autonomy with annatto plantation and processing to extract the dye, through financial and technical support, without commitment to exclusive purchase by Aveda." Aveda wanted "to be able to buy a product grown in an ecologically sustainable way and to produce a type of cosmetic with a highly differentiated value" (Bona, 1995, p. 3). The contract included Aveda financial support for the plantation, field activities, machine installation, and infrastructure at the village, in order to guarantee the proper processing and storage of seeds. Technical assistance was also included in the agreement, to be arranged by May Waddington, Aveda's project coordinator and manager in Brazil.

The annatto plantation was to include 13,000 shrubs, in combination with 5,300 peach palm trees (*Bactris gasipaes*), 3,800 Brazil nut trees (*Bertholletia excelsa*), 2,900 *guaraná* plants (*Paullinia cupana*), and some acerola plants (*Guilielma gasipaes*), all grown from seed and planted over a period of two years on a 35-hectare plot. Aveda financed a sum of US$49,600 for the project, which consisted of six parcels delivered directly to OAEYRG in 1993 and 1994, half as a grant and half as a loan. This money was used for planting, buying tools and machines, technical assistance, and transportation. Aveda reserved the right to make two annual appraisal visits to the area, and OAEYRG agreed to prepare budget reports for each installment of project funds, with new installments depending on approval of the previous budget and the ongoing results (Waddington, 1995). According to a preliminary report, production of annatto was expected to increase from 13.8 tons in the second year to 79 tons in the sixth year of the project (Waddington, 1995).

Such were the background and the goals of the project, negotiated and designed to fulfill the interests of two quite distinct partners: an indigenous group from Amazonia and an American corporation. The next two sections will examine how this project worked in its first years.

METHODOLOGY

Proper procedures and indicators for jointly evaluating ecological, economic, social, and cultural variables are crucial to appraise if a project contributes to sustainability. The successful completion of a specific task does not necessarily mean that a project aids in the sustainability of a community or an ecosystem. Thus, it is important to choose appropriate "pathways" and adequate evaluation and feedback procedures. Evaluation is also an important way to motivate people involved in the project to continue their contributions (Kline, 1995).

The next two sections cover crucial questions about a community-level project that combines the four dimensions of the concept of sustainability as discussed by Kline (1995): economic security, ecological integrity, quality of life, and empowerment with responsibility. Although no single community scores high in all four dimensions, this methodology allows one to evaluate how a community deals with change over time. According to Kline, a "pathway" defines the parameters to be measured in each dimension, deriving appropriate indicators through a process of community engagement. Due to time constraints, indicators were not derived for the Yawanawá tribe through a community engagement process. Hence, the assessment of the degree to which the Yawanawá–Aveda Bixa Project leads to progress toward sustainability was achieved through a framework developed by Alan AtKisson & Associates, Inc.

This qualitative methodology is a systematic evaluation based on the interpretation of available knowledge about a specific situation and is useful for analyzing the key strengths and weaknesses of sustainability goals. Created for communities in developed countries, this framework was devised originally for workers of the Washington State government responsible for implementing growth management legislation and monitoring comprehensive plans. It amounts to a checklist system that allows a comparison between various programs and projects seeking sustainability (AtKisson and LaFond, 1994). Although time constraints allowed only one field trip, ideally the information obtained in this manner should be compared and evaluated at two different periods of time or in relation to other projects.

Even though this study deals with an Indian tribal community, with values distinct from those of a community in a developed country, the framework was easily adapted because it covers the whole range of concerns expressed in the current literature about sustainable development in the Amazon. AtKisson's framework deals with three major topics, each of which integrates social, economic, cultural, and environmental concerns: overall contribution to sustainability (**S**), level of institutionalization (**I**), and degree of comprehensiveness and integration (**C**). For each of these topics a list of indicators is numerically scored, after a qualitative analysis. However, since the purpose of this study is not to compare different projects, indicators were not numerically scored. Instead, each indicator was analyzed qualitatively as "high," "medium," "low," or "too early to determine" in terms of its contribution to a sustainable pathway. Table 4.1 summarizes the major findings by showing each indicator and its chance of success in terms of being "high," "medium," "low," or "too early to determine." The approach takes the form of a case study, with a comprehensive and qualitative research strategy that relies on multiple sources of evidence, such as personal interviews, field observations, and documentary sources.[5]

ANALYSIS AND EVALUATION

Sustainability (S) *is the extent to which the project promotes long-term cultural, economic, and environmental health. The "S" scale measures if the project is in fact promoting positive steps toward sustainability.*

1. **The project has a long-term perspective (minimum 20 years).** *The time horizon of the project considers future generations.*

This indicator was scored with a medium chance of success. Annatto plants may live 50 years, but they have a commercial life of 30 years (Baliani, 1992).

Table 4.1 Sustainability Assessment of the Yawanawá–Aveda Bixa Project

	Likelihood that the Project Will Lead to a Sustainable Pathway			
Indicator	High	Medium	Low	Too Early to Determine
Sustainability				
1. Has a long-term perspective		X		
2. Maintains or restores ecosystem health	X			
3. Enhances economic vitality				X
4. Promotes equity and values diversity				X
5. Increases the resiliency of human and natural systems (includes culture, health, and education)	Cultural		Education Health	
6. Promotes cyclic use of resources	X			
7. Does not introduce new types of waste		X		
8. Empowers people and builds understanding		X		
9. Improves quality of life and individual sense of fulfillment		X		
Institutionalization				
10. Supported by population affected	X			
11. Generally understood by most people		X		
12. Supported by key interested groups			X	
13. Has moved from theory to practice				X
14. Has effective implementation mechanisms			X	
15. Has appropriate media attention				X
16. Is supported by a force of law or ethical standard that ensures its implementation		X		
17. Adequately designed, staffed, and financed in order to meet its professed goals			X	
Comprehensiveness and Integration				
18. Incorporates understanding of ecological, economic, social, and cultural issues		X		
19. Well integrated externally with other related efforts, initiatives, and programs		X		

New plantations can be made, of course. Interviews showed that community expectations are high in terms of the duration of the project. However, the situation is unclear regarding Aveda's support (finding market connections and providing technical capability) to produce and deliver the product.

2. **The project maintains or restores ecosystem health.** *It maintains overall ecological integrity and does not increase depletion of local natural resources.*

This indicator was scored with a high chance of success. From field observations, there is no evidence of depletion of natural resources. According to Bona's organic certification report, plantations were established at two previously deforested sites: 11 hectares at Sete Estrelas, the Katukina village, and 17 hectares around Nova Esperança, the Yawanawá village (Bona, 1995). No further deforestation was recorded as a result of the project. At the Yawanawá village, the 17 hectares planted with good quality annatto, some peach-palm (*Bactris gasipaes*), *acerola* (*Guilielma gasipaes*), *guaraná* (*Paullinia cupana*), and Brazil nut trees (*Bertholletia excelsa*) were established in second-growth areas called *capoeira*. In Bona's (1995) report, he states:

> The plantation in the community Yawanawá is the biggest and best cared for in the State of Acre. The vigor of the annatto shrubs is a reason of great happiness for the Yawanawá, especially because it is a product that has great potential in the regional, national and international markets. There is no doubt about the organic quality of the production, even though there is a lack of complementary efforts to store production after harvesting (p. 7).

3. **The project maintains or enhances economic vitality.** *It generates or supports sufficient economic activity to provide the tribe members with the goods they need.*

It is too early to determine whether the project is enhancing the economic vitality of the Yawanawá Indian tribe. Up to now, it has not, despite the good prospects of annatto in the national and international markets. According to preliminary report on the project, the Yawanawá expected to sell a production of 27.6 tons in 1995/96. This output, at the lowest expected price of US$2.00 per kilogram, would have given them a gross income of at least US$55,000. According to Biraci Brasil, the Yawanawá sold 6 tons of *colorau*[6] and lost 28 tons of annatto seeds due to problems with storage and transportation and to lack of buyers. Aveda still does not buy annatto straight from the Yawanawá. According to the original agreement, Aveda would connect them with a company that produces bixin, and Aveda would buy the bixin from them to use it as a raw material for its natural colors lipstick. However, logistics and infrastructure are still inadequate. The industrial processing of annatto seeds (to extract bixin) by a third party is also a limitation to the revival of the local economy.

4. **The project promotes equity and values diversity.** *The project itself is inclusive regarding economic and social elements.*

It is too early to determine the chance of success of this indicator. Due to lack of data, it is not possible to evaluate the distribution of project benefits and costs (monetary and non-monetary) within the Yawanawá community. This issue of equity is directly linked to the debate over how indigenous people might be integrated in the national and international market economy. Projects that involve indigenous peoples often bring up the issue of whether Indians should "await destruction" or actively "embrace capitalism." Indigenous peoples have a tradition of strong kinship and community ties. While these important connections may not be as strong as before in the Yawanawá community, they are still prevalent. Therefore, it is important to determine how dealing with money and producing for exchange impacts the community and how the community and its leaders deal with these changes.

Sales, acquisitions, and payments are managed both by OAEYRG coordinators (who distribute the merchandise bought to supply the village cooperative) and by the heads of the families. There is a potential for disagreement among community members about the management of project funds. On the other hand, there were recorded complaints about the fact that Aveda representatives would bring "presents" to only a few members of the tribe. Tribal members suggested that Aveda people should bring things that are needed by all, such as rubber boots, clothes, and working utensils.

In leader Biraci's words:

> What we do is a distribution based on solidarity, according to necessity, behavior and participation of each one in the village. Also, not everyone receives the same treatment. Some people help in organization or in an activity once a week. This is different from someone who works six times a week. In this way the coordinators must have a lot of sensibility and responsibility to make this distribution according to the needs of everyone that belongs to OAEYRG. This is done in accordance to the decisions of the directors of OAEYRG. The elders participate in the work with their knowledge, teaching traditional habits, telling stories, teaching about medicinal plants and organizing the *mariri* [traditional Yawanawá ritual]. So they contribute in the same way and have the same respect and love, just like the others (personal communication, March 13, 1996).

Despite community ties, access to the project's budget was found to be limited. Also, the method for revenue distribution is not clear. During field research this problem was being discussed among the Yawanawá, and a meeting was scheduled to discuss it.

5. The project increases the resilience of human and natural systems. *"Resilience" means the ability to adapt to changing or adverse circumstances.*

Even though the project is ecologically healthy and the ecosystem is not degraded, up to the date of the field research, the chance of success of this indicator was scored as high regarding cultural impacts, but low regarding education and health. Since the annatto plantation area had previously been deforested, and annatto is a native plant in the region, the analysis of this indicator focused more on the ability of human systems to adapt to changes, rather than natural systems. For this indicator, the main questions are how much the tribal community must change its habits because of the project and how much outside help it receives to make these changes.

Since the beginning of the project there has been a noticeable revival of the Yawanawá's cultural practices, such as song, rituals, religious sacraments with plants, games, and oral narration of their stories and myths, because most of the tribe is now reunited in a new place (after the project). By staying together, they are reinforcing their identity. Production of arts and crafts (arrows, baskets, pottery, and combs) was very limited before the project. It also went through a revival, but now it has almost ceased as most Yawanawá have engaged themselves in building the new village and in the annatto and the vegetable leather projects. Regarding education, the Yawanawá changed their practices since the project began. Previously, they had a school in Caxinawá, their old village. In the current village, the new school (built with funding of US$5,000 obtained through Rainforest Action Network) is being used to house the equipment to dry annatto seeds and the machinery used to extract the seeds from pods and manufacture *colorau*.

Regarding health issues, despite some improvements, the main problems still persist. Therefore, the chance of success of this indicator is still very low. Aveda donated a microscope to the village health post in order to help identify malaria infections, so that the proper medicine would be taken. In addition, Aveda arranged for a doctor to visit during the same period of this study's field research. However, medicine is not always available. The persistent health problems among the Yawanawá and the consequent process of seeking solutions inspired them to create "Living Pharmacy," a project that incorporates traditional herbal knowledge of shamans and elders and lessens dependency on modern allopathic medicine. This project is still seeking funding through partnerships with indigenous entities, humanitarian organizations, environmentalists, and private businesses concerned with indigenous and environmental issues.

6. **The project promotes cyclic use of resources.** *It reduces consumption of non-renewable resources and increases regenerative use of renewable ones.*

This indicator was scored with a high chance of success. The processing of the annatto seeds to make *colorau* uses renewable resources, except for the two machines that require a low-capacity diesel-powered generator.

7. **The project promotes waste reduction.** *It does not introduce new types of waste that will pollute the environment.*

This indicator was scored with a medium chance of success because currently it is not a major problem. The project itself does not introduce any type of chemical or solid waste. However, there is an indirect effect: with more visitors and industrialized goods circulating in the village, garbage has increased. The village lacks facilities for recycling or disposing of such waste.

8. **The project encourages involvement of all affected people in the decision-making processes.** *It empowers people and builds understanding beyond the leaders of the tribe. It does not break down existing community decision-making processes.*

This indicator was scored with a medium chance of success. A sustainable community should enable people to feel empowered and take responsibility based on a shared vision. On the one hand, there is not much evidence that the Yawanawá are empowered and independent in solving the problems related to project implementation. There are problems such as lack of training for operating and maintaining machines, shortage of basic implements and tools (wheelbarrows, bags, pans, and rubber boots), and technical difficulties in storage and transportation. On the other hand, the project stimulated discussions that went beyond the circle of tribal leaders, generating new insights about dealing with the outside world.

9. **The project improves quality of life and individual sense of fulfillment.** *Most affected people would report feeling that the project improves their lives.*

This indicator was scored with a medium chance of success. Despite the fact that the majority of the Indians feels that the project helped "open their eyes" and unify them, about 45% of respondents say that the project "still has a long way to go" in terms of improving their lives. The most important benefit of the project is the unification of the tribe in a new site, Nova Esperança.

Institutionalization (I) *is the extent to which the project is firmly embedded in civic life. The I-scale measures the distance between idea and reality. It assesses whether a project has social and political currency and what the mechanisms are to make it real.*

10. **The project is supported by the affected population.** *"Affected population" means those whose lives will experience direct impact as a result of the project.*

This indicator was scored with a high chance of success. The majority of the Indians support the project and any other project that allows economic returns while maintaining their traditional culture and their autonomy.

11. **The project is generally understood by most people.** *The majority understands what the project is trying to accomplish and why.*

This indicator was scored with a medium chance of success. Although the majority of the Yawanawá support the project, a large percentage of Indians interviewed could not describe exactly what the project is trying to accomplish and why. Although there is a great amount of happiness in terms of unifying the tribe and revitalizing its traditional culture, the project is still in its beginning stages, and the majority is not entirely aware of its complexity nor of their roles in making it work. For this to occur, they need more information and technical assistance.

12. **The project is supported by key interest groups.** *This includes businesses involved.*

This indicator was scored with a low chance of success. If the Yawanawá had the necessary support (by Aveda and other parties) in locating potential buyers on a national and international level, maybe they would not have lost 28 tons of annatto seeds in 1996.

13. The project has moved from theory to practice. *The project is not stuck in abstractions and promotes specific actions.*

It is too early to evaluate this indicator. The original contract has not been entirely implemented. Also, because the village is very remote, transportation has been a limiting factor.

14. The project has effective implementation mechanisms. *The project's administration, strategy, and tactics are effective.*

This indicator was scored as low. Project implementation depends on an active and permanent involvement between Aveda's project coordinator and the Yawanawá. As Aveda's liaison lives in Rio de Janeiro, timely response to eventual problems is difficult.

15. The project receives appropriate media attention. *"Appropriate" means the kind and amount of media necessary to ensure successful implementation.*

This indicator was scored as "too early to determine" because Aveda's use of the Yawanawá's name for marketing has not yet brought returns to the community. Also, spontaneous media attention has been very limited. An indigenous community does not necessarily need to receive media attention in order to be on the pathway to sustainability. On the one hand, if it were not for Aveda's marketing strategy, there would be no business partnership with a community in such a remote area. On the other hand, the fact that the project is supported by Aveda's marketing strategy does not necessarily ensure its successful implementation. So far, after three years of project implementation, the market of Uruku Lip Colors and Lip Sheers seems to fit well the needs of environmentally aware female U.S. consumers who do not wear lipstick solely out of vanity, but also because "it will help the Yawanawá Indians of the Brazilian rain forest revitalize their economy." The rationale in this approach is that "women can look pretty and do good at the same time simply by applying lipstick" (Prager, 1995). The Aveda Company flier that comes with every Uruku Lip Colors and Lip Sheers states:

> We have established a unique business partnership with the indigenous Yawanawá tribe in the Brazilian rain forest to grow the traditional uruku, scientifically known as *Bixa orellana* L. plant, which supplies natural pigment for all our Uruku Lip Colours and Lip Sheers. This unprecedented trade agreement not only supports the economy, autonomy and traditional cultures of the Yawanawá, but it helps preserve the rain forest — integral to all life on our planet. The uruku trees are planted in deforested areas to revitalize land previously stripped of its natural resources, and grown without petrochemical pesticides or fertilizers — which benefits the Yawanawá, the Earth and you.

It is important to recall that annatto shrubs can be planted nearly anywhere in the tropics. In order to produce the lipsticks, Aveda does not need to finance an annatto plantation in a remote area. However, Aveda chose to use raw botanical material grown organically in an indigenous community. Aveda's ecological mission statement includes the following principle: "to respect all ecological communities with which we interact, and to encourage small, local indigenous cultures and economies in the generation of materials for our products" (Aveda Corporation, 1996). This adds value to the product. However, the question of who is capturing this added value remains.

16. **The project is supported by force of law, regulation, or ethical standards.** *Some genuine center of authority is ensuring its implementation.*

This indicator was scored with a medium chance of success. The Yawanawá have their own ethics based on their trust of leader Biraci Brasil, respect for the elders, and the community decision-making process of OAEYRG. Everyone who lives in the community belongs to the OAEYRG and can speak up. The Yawanawá meet regularly, write minutes, and discuss all points of view. On Aveda's side, it may be too early to see how things will unfold. There are some questions as to whether it is really in stride with the original plan. Since the project started, there have been many unfulfilled expectations. Despite some inconsistencies, Aveda is still working with the Yawanawá and continues to support the Bixa Project. As for governmental regulatory powers, the Brazilian government has neither hindered nor supported the project. Given the problematic nature of indigenous policies in general, the absence of FUNAI is probably a positive variable.

17. **Scope and resources of the project relate well to scale and time frame.** *The project is adequately designed, staffed, and financed in order to meet its professed goals.*

This indicator was scored as low. First of all, community access to the project budget was limited. Therefore, it is difficult to evaluate if the project has adequate resources. Second, there are difficulties with the rhythms of the Yawanawá's region and their work, particularly with respect to transportation and health problems. Seed storage and transportation are serious problems. During the rainy season, annatto seeds are attacked by mold, and during the short dry season, transportation collapses. There are, therefore, technical and logistical problems in the project design.

Comprehensiveness and integration (C) *is the extent to which the project reflects awareness of linkages among its elements and the relationship between itself and the larger whole. The C-scale attempts to assess how well the project incorporates, in practical terms, the idea that "everything is tied to everything else."*

18. **The project is comprehensive with regard to ecological, economic, social, and cultural issues.** *While it may not focus on all four areas, the project incorporates an understanding of how they relate to one another.*

This indicator was scored with a medium chance of success. The project relates well to ecological and cultural issues, but not as well to economic and social issues. Despite Aveda's effort in making sure that annatto is grown without petrochemical pesticides or fertilizers, and the fact that the project is bringing about a cultural revival of the Yawanawá tribe, data gathered so far do not show an awareness of the linkages, especially regarding social and economic issues. Although it is not Aveda's responsibility to solve all the social and economic problems of the tribe, it is important for both the Indians and Aveda to consider the larger picture.

19. The project is well integrated externally with other related efforts, initiatives, and programs. *There is a demonstrated understanding of who else is doing similar work and a clear attempt to mesh well with other efforts.*

This indicator was scored with a medium chance of success. Biraci Brasil reported that some Yawanawá visited (with Aveda's support) the Kaiowá indigenous tribe, located in the state of Mato Grosso, where Aveda is investing in another project. The Kaiowá have been suffering much more cultural disruption and environmental degradation than the Yawanawá. The visit helped make the Yawanawá aware of the degradation they could cause if they insist on logging projects. Additionally, it was important for the Yawanawá Indians to share ideas with the Kaiowá about their experiences in finding economic alternatives that can be sustainable in the long run. This trip also helped to make the Yawanawá appreciate the richness of their environment compared to that of the Kaiowá homeland:

> I think that it is important to exchange experiences between communities so that we can learn how to develop our community, and I hope we can help them too. I hope that in the exchange of ideas they will receive many similar things that this project is bringing us and that it will help them develop and have a better life. And the same thing for us. We hope to exchange ideas for knowledge and work. I think this is what holds people together and gives strength to keep working (Tica, Yawanawá informant, March 13, 1996).

However, this visit alone did not change the Yawanawá's sense of isolation in terms of connecting with other organizations involved in similar work or potential customers outside Aveda's realm. Even though Biraci Brasil is aware of similar projects in the region,[7] there were no responses among the Yawanawá indicating awareness of, or communication with, other communities involved in similar projects. It remains to be seen whether Aveda's assurance of connecting the Yawanawá with other markets, including the United States, will be carried out.

Considering key indicators from the topic of sustainability (S), the conclusion is that the project scores: (1) high regarding ecological and cultural issues, (2) medium in terms of empowerment, and (3) low and uncertain regarding economic security and social well-being. Examining key indicators in the topic of institutionalization (**I**), the conclusion is that the project is supported by the majority of the Yawanawá, but it rates low and uncertain in regard to its mechanisms of accountability and implementation (planning, connections with potential customers, and technical assistance). Both indicators from the topic of comprehensiveness and integration (**C**) were scored medium, with the project only beginning to move in these directions.

Overall, the project is strong in relation to ecosystem integrity and cultural revival. However, it is suffering from logistical and technical difficulties, in addition to a certain degree of isolation, all of which are to be expected from its geographical location and from having its social basis in a relatively small and disenfranchised indigenous community. The most serious setback, however, is the project's low degree of economic reliability, caused by at least three interconnected circumstances: (1) lack of adequate seed storage and industrial processing facilities (that would require considerable input of resources), (2) logistical barriers (great distances and seasonality of water transportation), and (3) unmapped consumer markets.

CONCLUSIONS AND FOLLOW-UP

The previous sections examined the complexities of a direct business partnership between an American company and Brazil's Yawanawá Indian tribe. The emergence of green markets for rain forest products and the expansion of linkages between these markets and indigenous communities are attractive ideas for both people and institutions dedicated to forest conservation efforts and to the strengthening of the economic bases of those communities. However, indigenous groups will not enter a sustainable pathway even with well-designed, short-term projects if long-term structural problems concerning basic rights and needs (entitled homelands, political recognition, cultural integrity, education, health, and autonomy) are not addressed simultaneously. Obviously, other collective actors besides the community and the company must intervene. The ideal would be for Aveda to keep acting as a "catalyst" that seeks partnerships to work in collaboration with other institutions (such as universities and other governmental research bodies), especially during implementation of the industrial processing of annatto seeds at the village.

As the Yawanawá have a long history of contact with non-Indians, they have developed needs that cannot be satisfied entirely within their subsistence systems. Therefore, they must produce goods that generate monetary income. After unsuccessful attempts to do so without disrupting their culture, the contract with Aveda seemed like a fresh alternative of combining sustainable activities with cultural integrity. Aveda made a commitment to encourage small indigenous cultures and economies and to respect ecological integrity in its purchase of raw materials. Aveda's supplies of bixin currently do not come from the Yawanawá, and the company is free to continue buying bixin elsewhere. However, its "totally natural lip color" lipsticks are marketed with the Yawanawá's name and with its ethical stance of supporting a plantation organically farmed by indigenous people of the Amazon rain forest.

This study suggests that, to date, the project generated a thrust toward sustainability, but that it may not be sustainable. So far, the most evident benefit of the project is the unification of the tribe in a new place, allowing a cultural revival and reconnecting the Yawanawá with their traditional knowledge. This is not a small or easy gain. Although some indicators were scored with a high chance of success, especially concerning ecological and cultural factors, the social and economic issues that scored low or uncertain, if not corrected, are most likely to make the project fail, affecting, in the long run, indicators with high chances of success.

Adequate storage of perishable seeds, reliable transportation, and more diversified markets must be actively sought, especially because their joint absence is so common in low-technology economic activities located in frontier areas. There is much uncertainty about how the Yawanawá will preserve the quality of the final product and deliver it in a timely manner. According to the original agreement, Aveda would provide administrative support to the Yawanawá and help them network with other potential buyers. It is vital that the tribe connect with reliable customers.

Other economic problems should be mentioned in passing. The legality of commercial use of the Indians' name is vague and the contract says nothing specific about this matter. Uncertainty remains regarding the consistency of Aveda's involvement in the future. There has been a lack of technological assistance in training (machine setup, maintenance, and repair), storage, and packaging.

Some of the project's problems and issues in the social and cultural realms include the following questions: How much of the generated income is returning to the community in the form of goods needed, especially for healthcare and education? Will the project engender Yawanawá dependency on Aveda? Can the community manage activities from other projects so that they do not overlap with the climatic time frame and worker availability for annatto crops? Will the school building and the teachers' time continue to be used for the project? How can more people in the community understand why they are engaged in the project and how it will help them reach their own goals?

The cultural differences between the community of interest (Aveda) and the community of place (the Yawanawá), in the terms of Flora (1996), can present problems as well. Inevitably, there is an enormous cultural difference between the two parties in terms of rhythm, production, space, time, ownership, wealth, social structure, and relationship with nature. In order to be successful in creating sustainable pathways for the Yawanawá community, the project would benefit from more frequent interaction between the two communities in order to digest and facilitate the necessary flow of technical, scientific, and cultural information.

An orthodox application of the evaluation methodology would recommend a second full round of field research, to allow a comparison between two different moments (or in relation to other projects) and to derive more appropriate indicators through a community engagement process with the Yawanawá. A single round of field work was done in March of 1996. Follow-up information used herein was obtained only through contacts with individual Yawanawá leaders. In September of 1996, a Yawanawá tribal member traveled to the United States to discuss and update the original contract. According to Biraci Brasil (personal communication, October 15, 1996), the debt part of the first funding was converted into a donation, and the Yawanawá would soon receive a second installment of US$43,000. This money was to be invested in installing machinery to extract bixin at the village and to increase solar energy supply. Negotiations about Aveda's use of the Yawanawá name in marketing were postponed. At least Biraci Brasil was still unsatisfied with the overall results of the project (personal communication, May 3, 1997). His hopes were focused on the imminent extraction of bixin at the village. Bixin is a processed good that aggregates more value and is therefore expected to increase community revenues. Biraci Brasil stated that the Yawanawá want to "know more about the market" and to sell their product "only to people who know where it comes from." According to him, the community gained in cohesion and now wants to have a clearer view of its business partners.

The findings of this study may be of interest to other communities and businesses engaged in similar projects and negotiations. It is hoped that these findings will increase awareness of the issues raised, stimulate new partnerships, and aid in the development of effective projects that empower local communities while preserving their ecosystems. In other words, the sustainability of any such projects depends on the internal cohesiveness and the managerial capacity of each community, as well as the responsiveness of the business partners.

AUTHORS' NOTE

This chapter is a synthesis of a thesis elaborated by Sandra De Carlo for the Master's in Environmental Studies degree at The Evergreen State College, Olympia, Washington during 1994–96. José Drummond was a member of the thesis committee. Appreciation and thanks go to the members of the Yawanawá indigenous community and its main leader, Biraci Brasil, for sharing their lives and experiences on the Aveda–Bixa Project.

REFERENCES

Arnt, R. (1994). Perspectivas de futuro: Biotecnologia e direitos indígenas. In *FASE/FAOR — Forum da Amazônia Oriental* (pp. 9–19). Belém: Diversidade Ecossocial e Estratégias de Cooperação entre ONG's na Amazônia.

AtKisson, A. and LaFond, M. (1994). *Assessing sustainability projects — A prototype rating for comparative evaluation.* Seattle, WA: AtKisson & Associates Sustainable Development Series (available from 1 Kearny Street, 4th Floor, San Francisco, CA 94108; e-mail: AtKisson@Rprogress.org).

Aveda Corporation (1996). *Ecological mission statement* (Environmental Document, 01/27/95). Minneapolis, MN: Author (http://www.aveda.com).

Baliani, A. (1992). Corante natural para a indústria. In *A Lavoura* (pp. 19–22). Sociedade Nacional de Agricultura, January/February.

Bona, L. (1995). *Projeto de Plantio comercial de urucum (Bixa orellana L.). Organização dos Agricultores Extrativistas Yawanawá do Rio Gregório. Relatório de Inspeção* (unpublished report provided by author).

Carvalho, P. and Hein, M. (1989). Urucum — uma fonte de corante natural. In *Coletânea do ITAL — Instituto de Tecnologia de Alimentos (19*(1), 25–34). Campinas: ITAL, Secretaria de Agricultura e Abastecimento, Governo do Estado de São Paulo.

Comissão Pró-Índio do Acre (1996). *Breve histórico dos processos de organização dos Povos Yawanawá e Katukina do Rio Gregório.* Rio Branco, Acre: Author (unpublished paper).

Crespo, S. and Leitão, P. (1993). *O que o Brasileiro pensa da ecologia.* Rio de Janeiro: Agencia do Estado and Instituto de Estudos da Religião (ISER).

De Carlo, S. (1996). *A sustainability assessment of the Yawanawá–Aveda Bixa Project in the Brazilian Amazon Region.* Unpublished master's thesis. Olympia, WA: The Evergreen State College.

Flora, C. (1996). *Vital communities: Combining environmental and social capital.* Iowa City: Iowa State University (unpublished paper available from North Central Regional Center for Rural Development, Iowa State University).

Hecht S. and Cockburn, A. (1990). *The fate of the forest — developers, destroyers and defenders of the Amazon.* New York: HarperCollins.

Kitamura, P. (1994). *A Amazônia e o Desenvolvimento Sustentável.* Brasilia: Empresa Brasileira de Pesquisa Agropecuária — EMBRAPA-SPI.

Kline, E. (1995). *Sustainable community indicators.* Medford, MA: Global Development & Environment Institute, Tufts University.

Prager, E. (1995). The dressing room: Lipstick's just the half of it. *The New York Times,* June 25, p. 37.

SEPLAN — Secretaria de Planejamento do Estado (1993). *Programa estadual de zoneamento ecologico-econômico do Acre.* Rio Branco, Acre: Governo do Estado do Acre.

Smith, N., Williams, J.T., Plucknett, D., and Talbot, J. (1992). *Tropical forests and their crops.* Ithaca and London: Comstock Publishing Associates, Cornell University Press.

Waddington, M. (1995). *The Yawanawá–Aveda Bixa project.* Preliminary report.

Yawanawá, J. (1996). *Desenvolvimento econômico na comunidade Yawanawá.* Audiotape provided by Joaquim Yawanawá, Rio Branco, Acre.

ENDNOTES

1. Interview excerpted from Crespo and Leitão (1993, p. 59).

2. "The term *caboclo* refers to the Amazonian backwoodsman. Initially it was used to refer to detribalized Indians and various racial mixtures that included Indian blood" (Hecht and Cockburn, 1990, p. 279).

3. *Uruku* or *urucum* in Portuguese, or annatto in English (*Bixa orellana* L.), is a shrub 2 to 6 meters high. Its seeds contain bixin, a bright orange-red dye used by the Indians for decorative purposes and also as an insect repellent.

4. Annatto was a viable cash crop in Amazonia, and other parts of tropical America, until the advent of synthetic food colorants. After World War II, Red Dye No. 3 largely replaced annatto as a dye for the food and cosmetics industries. However, much concern arose in the 1970s, especially in developed countries, when the artificial dye was found to be carcinogenic. In early 1990, the U.S. Food and Drug Administration banned its use. This worldwide trend of replacing artificial food dyes with natural ones was reinforced by further stringent legislation in countries such as Canada, the United States, and Japan, increasing the demand for natural dyes. Annatto provides a wide range of color tones (from yellow through red to brown) depending on the process and action of different solvents used to extract the pigment bixin. Because this natural dye is safe for internal consumption and for skin applications, bixin is also being extensively used in industrialized countries by cosmetics companies for lipstick, suntan lotion, facial and hair lotions, and creams. In food processing, annatto is used to color butter and other dairy products, margarine, sausages, pastas, drinks, ice creams, liquors, tomato extract, juices, beers, and chocolates. Annatto is also used as a dye by the textile, furniture, and paint industries (Smith et al., 1992). Another use is suggested by Carvalho and Hein (1989): to achieve a desired color of egg yolks by adding annatto flour to chicken feed.

5. Field research was carried out during March 1996. Interviews were conducted with Biraci Brasil, the main leader of the Yawanawá; 27 residents of the Yawanawá community; Joaquim Yawanawá, who works with Pró-Indigenous Commission in Rio Branco, Acre; and May Waddington, Aveda's project coordinator in Brazil. In addition, conversations were held with Dionísio Soares (from the Acre Planning Secretariat and one of the designers of the Yawanawá–Aveda Bixa Project together with Biraci). Initial contact with both parties was provided by Beto Borges from the Rainforest Action Network, San Francisco, California. Additional questions sent to Aveda's U.S. headquarters in July 1996 were not answered. Biraci Brasil supplied additional information by telephone..

6. In Brazil, *colorau* (a mixture of 15% annatto seed flour and 85% corn flour) is popularly stirred into stews and rice dishes to add color.

7. For instance, the reforestation project also involves the use of agroforestry systems using deforested areas to grow perennial species like *cupuaçú* (*Theobroma grandiflorum*), peach-palm (*Bactris gasipaes*), and Brazil nut (*Bertholettia excelsa*), combined with mutually cooperative producer organizations, access to credit and markets, as well as technical and institutional assistance.

CHAPTER 5

Eco-Village Development:
A Report from Southern Appalachia

Jonathan M. Scherch

This chapter will introduce and discuss sustainable communities, referred to here as eco-villages, and the bioregional context with which they are interdependent. An in-depth look at the structure and accomplishments of one eco-village (Narrow Ridge Earth Literacy Center) will be presented, followed by parallel summaries of two additional eco-village examples (Earthaven Village and the Eco-Village Training Center at The Farm). Finally, the chapter will serve to offer an assessment of the contribution of eco-villages toward the goal of modeling ways and enabling people to live more sustainably within their own bioregion.

WHAT IS AN ECO-VILLAGE?

The definition of a sustainable community embodies and is strengthened by the conceptual work of various authors and thinkers. Thus, the concepts of sustainable communities, eco-villages, sustainable neighborhoods, and the like are treated as synonymous in meaning and mission. Developing a concept of community which is bioregionally appropriate is important since, as Haenke (1995) states,

> the ecosystem, the watershed, the bioregion: these are the context, the boundary, the basic foundations of community. Lest our communities be parasitic and unsustainable, rights equal to our human rights must be secured for all living things of the ecosystem community in which our human communities are embedded (p. 119).

With this in mind, Gilman's (1995) concept of an eco-village, defined as "a human-scale, full-featured settlement in which human activities are harmlessly

1-57444-129-9/98/$0.00+$.50
© 1998 by CRC Press LLC

integrated into the natural world in a way that is supportive of healthy human development and can be successfully continued into the indefinite future" (p. 109), seems consistent.

The Gaia Trust (1995) defines eco-villages to be:

> an integrated solution to the global social and ecological crisis, and are as appropriate to the industrialised world, both urban and rural, as to the remaining two thirds of the world. Eco-villages are in essence a modern attempt by human-kind to live in harmony with nature and with each other. They represent a "leading edge" in the movement towards developing sustainable human settle-ments and provide a testing ground for new ideas, techniques and technologies which can then be integrated into the mainstream (p. 1).

Parallel to this description, Gilman (1995) notes that:

> in an eco-village, or sustainable neighborhood, people can be close to where much of their food is grown and their livelihoods. The physical and economic environment is arranged so that quality time with family, friends, and community is possible. Leisure, recreational, and civic activities are within walking or short non-polluting commute distance (p. 109).

Arkin (1995) indicates that the "public interest purposes" (p. 112) of eco-villages are such that they serve to: (a) demonstrate low-impact, high-quality lifestyles; (b) reduce the burden of government by increasing neighborhood self-reliance; (c) reverse the negative environmental, social, and economic impacts of current growth and development practices; (d) model sustainable patterns of development in the industrialized world, especially the United States, to encour-age developing communities to bypass our unsustainable patterns; and (e) show-case the talents and skills of those working toward a sustainable future.

THINKING BIOREGIONALLY

Bioregionalism refers to a comprehensive new way of defining and under-standing the place where we live and living in that place sustainably and respectfully. "Bioregionalism speaks to the heart of community. If we are to continue living on Earth, the definition of community has to include all the living things in our ecosystem" (Haenke, 1995, p. 118). In support of this definition, John Nolt (personal communication, December 9, 1996) defines bioregionalism as an attempt to organize community, economic, and political action in units delineated by ecological systems and relationships. Thus, in short, thinking bioregionally reflects the acknowledgment of and a deliberate attempt to enrich the interdependent relationships between hu-man and non-human communities.

THE SOUTHERN APPALACHIAN BIOREGION: WHAT'S THERE?

For the present study, the Southern Appalachian bioregion is broadly defined by the use of three sources: (1) the watershed of the Tennessee River as identified by the Tennessee Valley Authority (1995), including East Tennessee; (2) the province called *Katúah,* the name derived by the Cherokee Nation for the watersheds above 1,200 feet (and associated coves) from the Northern Georgia mountains northeastward to near Roanoke, Virginia (Barnes, 1996); and (3) the south-central portions of the Southern Appalachian Assessment Area defined by the Southern Appalachian Man and the Biosphere (SAMAB) cooperative (SAMAB Cooperative, 1996). Together, these three references serve to form a conceptual map of a bioregion which includes several million people across six states, five national forests, and three national parks and preserves.

The presence of the temperate range of the Great Smoky Mountains National Park and the streams and tributaries feeding the French Broad, Holston, Pigeon, and Tennessee rivers, among other sources, deliver to the whole bioregion its lifeline. For centuries, the mountains and rivers of Southern Appalachia have provided the resources which have sustained both human and non-human communities. Old-growth forests, home to rare and endangered plants, animals, and countless unseen microorganisms (Western North Carolina Alliance Old Growth Committee, 1995), as well as vital systems, have collectively formed a web of life (Barnes and Messick, 1991). Unfortunately, like bioregions elsewhere, that of the Southern Appalachians is in crisis.

A BIOREGIONAL CRISIS OF ECONOMY

Essentially, the interaction between the spheres of economy and ecology represent what Harte (1993) remarks is the confluence of interest between a healthy human society and the well-being of natural ecosystems. In the Southern Appalachian bioregion, this interaction generally has not been a healthy one. More specifically, according to Wheeler (1991), this bioregion has endured the interaction of two types of economies: *a living economy* and a *human economy.* The *living economy* refers to the elemental powers "refined and individualized into acorns and whirling poplar seeds, luminous trilliums, insect larva crawling under stream rocks, a grouse thrumming in the twilight — beings that live and die, eat and are eaten, closely bound to the web of existence" (Wheeler, 1991, p. 1). The *human economy* refers to "an accounting of how we live within the greater (living) economy and utilize its energies to support our own existence" (p. 1). Thus, the acknowledgment of the presence and functioning of both economies provides a reference point for examining how the current behavior and direction of the bioregional human economy might well be engaged, albeit unknowingly, in the gradual compromise of the welfare of the living economy.

Recent research has served to document, in a complex and synergistic fashion, the full scale of social and environmental transactions which have altered the character and functioning of the bioregion. More specifically, the work of the Foundation for Global Sustainability, in its development of *What Have We Done? The Foundation for Global Sustainability's State of the Bioregion Report for the Upper Tennessee Valley and the Southern Appalachian Mountains* (Nolt et al., 1997), has documented the socio-environmental effects of various activities related to industrial development, rapid population growth, rural-to-urban land conversion, resource acquisition and consumption, and waste management issues, among many others.

Adequate understanding of the nature, scope, and effects of the issues facing the bioregion requires sophisticated studies which are capable of examining these issues for their individual and synergistic effects. While this type of study is difficult to conduct, one such effort produced the recent *Southern Appalachian Assessment* (SAA) (SAMAB Cooperative, 1996). The findings of the SAA (SAMAB Cooperative, 1996) revealed no major crises for the bioregion, but some findings were considered "worrisome" (p. 4). More specifically, the SAA noted problems caused by forest pests and ecological changes to the region's forests; pollution and acidity levels affecting water quality, fish species, and human uses in certain streams; and pressures on natural resources by human development and associated conflict about their usage. Though non-specified in its identification of the likely sources of these problems, thus rendering its findings politically benign, the SAA produced relevant information regarding the status and future of bioregional communities.

However such issues might be characterized, their effects are apparent, as described by Nolt (1996):

> But how different this land is today! Criss-crossed by roads and power lines; studded with microwave towers; noisy with the motors of cars, trucks, jet airliners, and chain saws; water and air polluted, the blue mist of the mountains now often a hazy sulfurous white; suburban sprawl blossoming outward from cities small and large; the forests cut and cut again, the big trees all but gone, the solitude unrecoverable; the rivers dammed, tamed, and silting up; the meadows plowed, planted, paved (p. 1).

Thus, while there may be semantic disagreement regarding the gravity of the issues facing the bioregion, the conclusion can be drawn that these issues are adversely affecting the health and welfare of the bioregion. Whether formally stated or not, this conclusion is beginning to mobilize energies to address, resolve, and avoid these issues.

A SUSTAINABLE RESPONSE

In many ways and for a variety of reasons, people living and working within the bioregion have chosen to respond to this crisis by pursuing innovative lifestyles and enterprises which, essentially, emphasize and prioritize taking responsibility for their actions and choices. More specifically, many people have chosen to reduce their consumption of commercial foods by either growing their own foods or investing in community-supported agriculture programs. Others have chosen to build their homes out of locally available, renewable resources, such as straw bales, cob, wood, or even scrap materials, so as to minimize the use of costly building supplies.

Still others have endeavored to become more energy independent, joining, both directly and indirectly, a larger movement in using various alternative so-called home power energy sources. The sources, including solar, wind, and micro-hydro power, are used to power homes and businesses. Thus, they serve to raise important questions about the operations of centralized, electric utilities, as well as some of the assumptions traditionally incorporated in energy policy analysis and decision making (Tatum, 1992).

Building on the concept of community of interest (Kahn, 1991; Weil, 1995), and in response to recent calls for expanded efforts at community building (Weil, 1996), the following examples of sustainable communities offer portraits of such innovative living. While it is impossible for this study to present a comprehensive report on the motivations for why these communities and their members came together, the most apparent and perhaps most important feature of these groups is their common interest in living more sustainably and responsibly.

A CASE EXAMPLE: NARROW RIDGE EARTH LITERACY CENTER

Founded in the late 1980s, the Narrow Ridge Earth Literacy Center (referred to hereafter as Narrow Ridge) is located in the Hogskin Valley approximately 45 miles north of Knoxville, Tennessee. According to Bill Nickle (1995), the founder and director, Narrow Ridge got its name from theologian Martin Buber's idea that the purpose of each of us is "to walk the narrow ridge between the inward and outward life" (p. 37). Buber believed that we must revere the sacred in other beings to overcome the alienation from one another and from the earth that so many people feel in modern culture. Thus, "the problem is not whether you slide off the narrow ridge, but whether you get back on" (Nickle, 1995, p. 37).

As an Earth Literacy Center, the mission of Narrow Ridge is to provide environmental programs and retreat activities based on community, sustainability, and spirituality, which are intertwined in Nickle's thinking (Kleihauer, 1996). The concept of earth literacy is defined by Smith (1996a) to be a budding curriculum aimed at freedom as well as an environmental movement in which life on earth is seen to be at a turning point. The turning point (Smith, 1996b) is a crisis in our perception of reality. To respond to that crisis, we are beginning to rethink the way we live our workaday lives. Noting that the shift to earth-literate living requires a radical change in perception, Smith (1996b) explains that earth literacy grew out of a communion agreement made when individuals from two farm-based learning centers, two colleges, and a university (Genesis Farm, Narrow Ridge, Southwestern College, Miami–Dade Community College, and St. Thomas University) met to build relationships that would enrich the earth in the bioregions where they lived.

The developing community at Narrow Ridge is one which prioritizes healthy and just relationships between the people who live and/or work there and the inhabitants of the surrounding natural environment.

> [T]he occupants of the ridges and valleys where we live are both human and non-humans. The humans come and go. The other creatures who make their homes here are the real inhabitants. Their very lives depend on the health of the earth, air, and water. We humans at Narrow Ridge are just beginning to learn again what it means to live with the land, to re-inhabit earth (Bill Nickle, personal communication, December 17, 1996).

Thus, Narrow Ridge serves as an important bioregional model of:

> ...how we interface with people and with earth and all of its biosphere. If what Narrow Ridge is about is to study, to teach, and to demonstrate what we think can be done in our particular locale, then maybe other people can get other ideas of what they also could do in the bioregion (Bill Nickle, personal communica-tion, December 17, 1996).

In this regard, the planning and development work of the Narrow Ridge community reflects the purposeful use of many alternative designs, methods, and techniques which simply demonstrate viable, different ways of living.

Settlement at Narrow Ridge

In advance of efforts to settle on the land of Narrow Ridge in the late 1980s, Sosadeeter (1993) conducted a site analysis using the McHarg Overlay Method (McHarg, 1971) to determine how human settlement could be undertaken with respect to the features and needs of the present natural systems. Soils, topogra-phy, hydrology, vehicle/pedestrian access, solar access, vegetation, climate, existing structures, and view-sheds, among others, were all identified, mapped, and evaluated (Sosadeeter, 1993). Accordingly, Sosadeeter (1993) numerically rated each of these elements for its natural suitability for housing and road development, annual and perennial crop production, pasture land, passive and active recreation, septic tank absorption fields, and woodlands and wildlife habitat. Such an analysis provided important preliminary information for the sustainable development of the land trust communities.

The Land Trust Communities

Narrow Ridge is comprised of three land trust communities (Hogskin, Black Fox, and Log Mountain), which taken together encompass 351 acres of land (105, 168, and 78 acres, respectively). Within these three areas, portions of the land have been reserved as wilderness areas (10, 108, and 38 acres, respectively). The 10 acres reserved for wilderness within the Hogskin land trust may also be used for gardening, although no trees may be cut.

Land trust homesites are leased; community members do not own the land. What a leaseholder may do on a particular site can vary from one land trust to another, but "there is a bottom line which Narrow Ridge has set, as far as conservation restrictions and easements are concerned" (Bill Nickle, personal communication, December 17, 1996). While the bylaws of the land trusts impose land use restrictions on leaseholder activities, such restrictions are viewed as being in keeping with the interests and expectations of the members of the community.

> A fundamental "bottom line" is that folks are really concerned about Earth...Some are hard-core environmentalists, and others are more "up-town" types. They all have a basic concern for Earth, and in each of the land trusts there is a recognition and an affirmation of what Narrow Ridge stands for and that they, in themselves, wish to abide by the spirit of Narrow Ridge in all that they do (Bill Nickle, personal communication, December 17, 1996).

Currently, four staff live at Narrow Ridge, including an educational administrator, a homeopathic practitioner, a cooper, and the director, all of whom are hired by a 12-member board of directors. The land trust communities are currently designed to accommodate 35 residential construction sites, in addition to the current residences of staff. At some point in the future, upwards of 180 people might be supported by the land trust's settlements.

An important feature which underpins the vision of Narrow Ridge is its articulated and observable commitment to racial and gender equality and the respectful affirmation of the value of home. In short, this commitment appears to be aimed at restoring the home as a cultural center of family life and serving as a principal source of healthy socialization and self-sufficient life skill development. This commitment comes in response to Bill Nickle's observation of how the home has changed over time (personal communication, December 17, 1996). No longer providing these benefits as it once had, the home is a dependent place of domestic work, including child care, which is largely devalued and inadequately supported by the larger society and performed predominantly by women. In contrast, at Narrow Ridge, child care and homemaking are viewed and affirmed as equally valuable in comparison to other duties and responsibilities within the community. Furthermore, a visitor to Narrow Ridge may see mainstream gender roles merged. For example, a farming couple might share the duties of tending to animals, planting and harvesting crops, and child care responsibilities, while other men and women might take turns stacking straw bales toward the building of a new structure and jointly holding positions of authority within organizational subcommittees.

Racial and gender equality notwithstanding, people come to Narrow Ridge for many different reasons. Some come with definitive purposes and plans with regard to transitioning onto their share of the land trust, while others may not have such a clear understanding of what their opportunity will become. "Some of the ties are very strong, and others of them are tenuous at this point. People just like the concept and say 'I'm not really sure about how I'm going to fit in.' They're not exactly sure how it's going to work, but they're committed to coming" (Bill Nickle, personal communication, December 17, 1996).

Examples of Sustainable, Earth-Friendly Living

As alluded to above, woven into the community fabric of Narrow Ridge is the imperative to live responsibly with regard to the acquisition and consumption of resources and the management of residual wastes. For example, in an effort to address issues related to the origin of utility-derived electricity and fossil fuels, Narrow Ridge has integrated various alternative energy systems into its building designs. Passive and active solar energies for producing electricity, lighting, and heat are currently in use at the Narrow Ridge Community Center, residential hermitages, and the recently erected student dormitory. These and other dwellings are designed to be earth-friendly with regard to the construction materials used, their energy efficiency rating, and use of composting toilets and natural gray-water drain fields.

Nonetheless, while efforts are made to minimize the use of utility-derived energy sources and fossil fuels, Narrow Ridge recognizes both limitations presented by the land (woodland areas with minimal solar access) as well as individual preferences (use of energy-inefficient appliances) which constrain, for now, living completely apart from central utility power sources. With this in mind, efforts are currently under way to explore the development of a community solar bank, which would support multiple homes from a central solar or wind-power source, as well as the use of alternative fuels (e.g., alcohols, biogas) for cooking, heating, and transportation.

Building construction has included the use of sustainable resources which can be acquired on-site or purchased locally in support of a local small business. For example, various buildings have incorporated the use of straw bale, cob, wood, and other techniques into the design and construction. These techniques offer various benefits such as (a) low-cost and participatory construction using

appropriate, locally derived, renewable resources; (b) exceptional energy efficiency and temperature control; and (c) resistance to hazards such as fire and
earthquake.

The Challenge of Building Community Diversity

As a rural community located within an hour's drive of Knoxville, Tennessee,
Narrow Ridge has developed relationships between residents and visitors that
serve to maintain important rural-to-urban linkages. The developing programs of
Narrow Ridge's organic farm are expected to build on these relationships,
eventually recruiting participation in its community-supported agriculture (CSA)
program, as well as its workshops, harvest celebrations, and so forth, as it draws
people from both rural and urban settings. The various educational programs to
be offered at Narrow Ridge, along with the availability of the unique hermitage-
style accommodations and student dormitory, also are anticipated to enlist the
participation of many people of all ages.

According to Bill Nickle (personal communication, December 17, 1996), the
Narrow Ridge Board of Directors has tried to support and promote diversity, as
it pertains to membership philosophy. Ethnicity, religious identity, and so forth
are treated as very important and thus are reflected in the interests, plans, and
spirit of the members: "There are folks that are a lot more intentional, using
straw bale or cob house constructions, while others are wanting to do all they
can but wish to have modern conveniences — the electric stove, living on the
grid" (Bill Nickle, personal communication, December 17, 1996). Nonetheless,
garnering the participation of minority groups and people of color has been
especially challenging for Narrow Ridge.

Perhaps similar to other communities, Narrow Ridge recognizes the necessary challenge of reaching out to new communities, including all minority
groups, in both rural and urban areas, in an effort to instill confidence and trust
in those whom it seeks to include in the community. Nonetheless, the white and
predominantly middle-class Narrow Ridge community has had limited success
in gaining the participation of minority groups and people of color. However,
efforts have been made to increase interaction with various groups and communities, including speaking engagements with community groups, schools, and
churches, as well as tours and events held at Narrow Ridge. Nonetheless, issues
of racial discrimination continue to mar the social landscape in the southern
United States, as evidenced in recent church burnings. Existing and/or perceived social and economic inequalities result in strained or simply distanced
relations across social groups which may be impeding the effectiveness of the
efforts.

Also, the trends of rural-to-urban migration, involving people in search of
employment, housing, and an improved standard of living for themselves and
their families, are now difficult to reverse. "Narrow Ridge is in a curious bind.

For many people, the cabin on the mountain is not where they wanted to be; that's where they were" (Bill Nickle, personal communication, December 17, 1996). Like other sustainable living initiatives, the individual and social benefits to be derived by Narrow Ridge's example have not yet caught the favor of the "mainstream" person; perhaps the short-term gains of mainstream living, albeit environmentally unsustainable, are yet too enticing when compared to those gains offered at Narrow Ridge.

Working Toward Economic Self-Sufficiency

Various entrepreneurial ventures are under way which are designed to help Narrow Ridge achieve and sustain economic self-sufficiency, while modeling viable socially responsible ventures and preserving a rich cultural heritage. For example, planning efforts are currently under way at the Narrow Ridge farm to begin producing organic vegetables, flowers, honey, and herbs, as part of a CSA program for individuals and families residing within and outside of the immediate Narrow Ridge community. Such a venture calls attention to the undesirable social and environmental effects of centralized and largely petrochemical-dependent agribusiness, while offering a new range of choices in the origin, production, and nutritional value of food.

Residents hope that the Narrow Ridge farm programs will eventually produce other resources, including organic compost, greenhouse building design and construction, and various agricultural workshops and tutorials. In short, the development of a CSA program will be a significant achievement for Narrow Ridge and the surrounding communities as "community-supported agriculture is a very viable agricultural way of looking at things, not just from an economic point of view, but culturally and every other way; it helps create the bonds of community" (Bill Nickle, personal communication, December 17, 1996).

In complement to the activities of the CSA, Narrow Ridge is developing ventures involving various skills and disciplines, including coopering and woodworking, blacksmithing, seamstressing and weaving, potting, and home-building design and construction. For these occupations to function, emphasis has been placed on using locally found resources acquired and used in sustainable ways. In addition, Narrow Ridge has created a publishing company, Earth Knows Publications, and some day hopes to have its own printing and binding operation. To date, the company has produced two book publications, including *Down to Earth: Toward a Philosophy of Nonviolent Living* (Nolt, 1995) and *What Have We Done? The Foundation for Global Sustainability's State of the Bioregion Report for the Upper Tennessee Valley and Southern Appalachian Mountains* (Nolt et al., 1997). In providing employment opportunities for those living within Narrow Ridge, as well as for those in the surrounding communities, these and other ventures are intended to function as sustainable sources of income, as ways of preserving and reintroducing traditional life skills across generations, and as the means by which persons of all ages can contribute to the healthy functioning of the community.

Preserving Bioregional Culture and History

The preservation of bioregional culture is an important element in the development and activities of the Narrow Ridge community. Craft-making, use of traditional manual tools and techniques, traditional food preparation techniques (including the use of a sorghum press in the future), and community events and celebrations, such as Hogskin History Day and annual solstice celebrations, generally reflect a belief that "the old ways of doing things are still good ways" (Bill Nickle, personal communication, December 17, 1996). Events like the Hogskin History Day "try to preserve the heritage that is there — what was school like, what was the religious experience like, what was the food like, homes, housing, music — all of those things we have tried to emphasize and pick-up on, and affirm" (Bill Nickle, personal communication, December 17, 1996).

Moreover, the preservation and reintroduction of the cultural history of the area are viewed as an important means of reconnection between current and past generations. Such a reconnection:

> is basic; it goes back to Earth literacy itself. In that, Earth teaches us what people have known for hundreds or thousands of years, and it has only been in the last 150 years that we've completely taken it off an "automatic pilot" and [taken] control. So, for us to perceive what has happened, forces us to look back. It's not that we're going to go back and live like our great grandparents did, but we need to have some sort of an understanding of the ways they lived, and why they lived like they did, in order to put appropriate technology and the old ways together (Bill Nickle, personal communication, December 17, 1996).

Educational Programs

Narrow Ridge aims to offer a wide range of educational programs designed to show and teach people about the viability and applicability of healthy, sustainable ways of living. The Narrow Ridge farm program offers apprenticeships for those interested in pursuing organic farming and CSA. Tours of the farm, and of the larger Narrow Ridge community, are also conducted for primary and secondary school classes, church and community groups, as well as scientists and researchers.

Narrow Ridge also offers a college-level Earth Literacy Internship program which is designed to "immerse the learner in a community of people committed to values of simplicity and engagement in the life and issues of the bioregion" (Narrow Ridge Earth Literacy Center [NRELC], 1995, p. 1). The program provides a three-credit course, on-site accommodations, and a comprehensive living-and-working experience which attends to the mission and activities of the Narrow Ridge community. Through the program opportunities exist for the intern to work with local organizations and individuals whose field of work and interests are shared by the student. Such opportunities might include business and agricultural cooperatives; social and health services; legal and academic research; social change and leadership training programs; historic preservation and museum programs; and individual entrepreneurs in music, theater, art, and crafts (NRELC, 1995).

Generally, in keeping with the tenets of earth literacy, this internship program and the accompanying seminars serve to "deepen the learner's understanding and sense of responsibility for the ways in which every profession, every vocation, is practiced" (NRELC, 1995, p. 1). More specifically, a major tenet of the program is reflected in its:

> ...attempts to provide a more meaningful context for the learner's work in the world, whatever that work might be. [The internship] reaffirms the spiritual context for all human vocations. The learner's chosen profession is perceived not in terms of its potential for economic "success" and status, but, rather, in terms of its potential for fulfillment and purpose (NRELC, 1995, p. 1).

Thus, students are exposed to and come to learn the value of sustainable, self-sufficient living and working in the context of earth literacy principles.

The purposeful mix of educational programs is designed to accommodate and include the varied interests and motivations of the visiting student, class, or group. Thus:

> People can come for strictly academic work, but then a person might come and say, "Hey, I don't want academic credit, I want to come and work for six weeks with you on a building, or maybe it's an apprenticeship with the farm or whatever. Other folks come for their own vision quest, and that's one reason why we have 108 acres of wilderness for persons, and they mostly are young adults, who want to spend two, three, four nights and days alone. We try to tailor a quest for the individual's interests and needs (Bill Nickle, personal communication, December 17, 1996).

Interaction with Other Sustainable Communities

The sharing of information about appropriate technologies and efforts at preserving land and establishing other land trust communities are examples of how Narrow Ridge interacts with other sustainable communities bioregionally, nationally, and internationally. In view of the challenges facing the inhabitants of the bioregion, the exchange of ideas, methods, techniques, and experiences across communities is valuable.

We're going to find out in the next fifteen to twenty years that there are going to be a whole lot more people that are going to be coming to grips with this point of trying to preserve lands. I like to hear what others are doing, and there are folks who are sharing ideas with us. If we don't start preserving lands, not only will we not have wilderness but we won't have farms to farm (Bill Nickle, personal communication, December 17, 1996).

Cross-community interaction will likely strengthen with increased visitation by and to Narrow Ridge, as will the preparation of and building within the land trust sites. "As more people begin to build their homes, they will be relating to other folks in other places that have built as well" (Bill Nickle, personal communication, December 17, 1996). For example, Narrow Ridge has successfully designed and built the first two-story straw bale structure (the student dormitory) in the bioregion and, consequently, has received many inquiries about the particulars on this project.

Two additional sustainable communities offer excellent examples of those with which Narrow Ridge residents might interact and learn from. Located on the fringe of the bioregion under consideration here, these communities share and emphasize similar value bases of sustainable, responsible living. The following sections briefly describe these two communities.

EARTHAVEN VILLAGE

Marsh (1996), an Earthaven member, said: "[This community] is envisioned as a permaculture-based eco-spiritual community that will learn and demonstrate the skills and technologies of a viable village culture appropriate to our historical moment and our bioregional context (the southern Appalachian mountains)" (p. 23). Located in the foothills of western North Carolina, the Earthaven community is integrating principles of permaculture, as coined by Mollison and Holmegren in the 1970s (Bane, 1996): appropriate alternative energy technologies such as solar and micro-hydro, eco-friendly housing design and construction including straw bale and cob construction techniques, and cooperative community living which emphasizes gender equality and encourages membership diversity.

Beginning with the physical, financial, and spiritual investments of some 12 members, the Earthaven community was founded in 1994. These investments are an important feature to the development of the community as:

Consistent with the ethics of self-reliance and the aim of demonstrating accessible alternatives to conventional development, the Earthaven community chose not to seek bank financing but instead to finance the project privately through the sale of leased site holdings and memberships, and the development of a member-owned investment cooperative referred to as Earthshares (Marsh, 1996, p. 24).

Thus, in retrospect, Marsh (1996) notes that "while this decision has limited our development capital somewhat, it has also freed us to build the village as we choose, while learning to make the most of the human and natural resources that we do have" (p. 24).

Thus, not surprisingly, many resulting features of the Earthaven model are easily discernible when compared to the conventional, mainstream community. For example, the community built and now uses a straw bale kitchen and bathhouses including composting toilets, which were designed for common use as a sustainable and responsible alternative to building individual in-home facilities. More generally, another discernible feature of the community, as da Silva (1996), an Earthaven member, notes, is that:

> [W]e (might) go beyond what Scott Peck called pseudo-community to real community. We're learning we don't have to be especially polite with one another, or expect good vibes all the time. We're learning individuals can be at significant odds without the community teetering. We're learning we can scrap a plan and keep our core vision (p. 6).

Like other intentional sustainable community initiatives, Earthaven has taken on the challenges of cultivating and assuming an alternative lifestyle.

> We want to reduce our dependence on the automobile and discourage commuter lifestyles by creating a viable local economy so that we can work where we live and live where we play. We plan to restore and care for our waters so that they leave our land cleaner than when they entered. We intend to restore biodiversity and health to our forests and to create a sanctuary for native, endangered, and useful plants and animals (Marsh, 1996, p. 23).

ECO-VILLAGE TRAINING CENTER AT THE FARM

Formed during the late 1960s and early 1970s in south-central Tennessee, The Farm is one of the largest and longest lasting communities formed during the so-called hippie movement (Traugot, 1994). In short, the 320 long-haired hippies who started The Farm came to Tennessee believing in clean air, healthy babies, honest work, non-violence, safe energy, cheap transportation, and rock and roll music (Bates, 1988)!

The Farm started as a spiritual community with a totally communal economy and then became a cooperative with a hybrid economy. At its peak, The Farm had nearly 1,500 residents. The Farm has been quite active and influential in the fields of vegetarian diet, midwifery and home birth, appropriate technology, clean and sustainable energy, overseas relief and development, environmental and human rights law, and in the study and promotion of intentional communities. Also, through Plenty, its charitable relief and development organization, The Farm has initiated and facilitated development projects all over the world (Traugot, 1994).

Bates (1988) documents the milestones of the evolving Farm community. He describes various achievements in, for example, the construction of solar housing and wind/solar alternative energy systems; communal food and agriculture production, with soy being the principal protein food; innovative modes of transportation, including hybrid electric vehicles; communication systems including ham and CB radio stations and an on-site telephone service referred to as Beatnik Bell; and electronic products such as radiation detectors.

Building on the history of The Farm, the Eco-Village Training Center (ETC) at The Farm currently provides training courses in permaculture design and installation, alternative energy technologies, affordable housing design, and sustainable community development, among many other programs. These activities follow a basic belief which underpins ETC educational programs: "to make the transition toward a sustainable society it is imperative that we take responsibility for our own lives and meet our basic needs for food, shelter, energy, gainful employment, and supportive community" (ETC, 1996, p. 2).

The ETC was created after The Farm was chosen as one of a seed group of sustainable communities for the developing Global Eco-Village Network (GEN). These communities were chosen for a variety of reasons, including geographical spread, attractiveness as models, ecological and spiritual awareness, and personal contacts. Other member communities of GEN are located in Australia, Denmark, Hungary, India, Germany, Russia, Scotland, and the United States (GEN, 1996).

SIGNIFICANCE OF EXPERIMENTAL ECO-VILLAGES

Essentially, the livelihoods of those who live and work within these intentional sustainable communities seem devoted to achieving balance between a desirable quality of life standard and the means by which such a standard is defined, attained, and sustained. While the people who comprise these three communities are not easily grouped or characterized, their efforts at living more responsibly appear to serve as a worthwhile point of commonality. More specifically, their example of living constitutes a sophisticated social critique. Working and crafting a common sustainable vision, and their courage and resilience to the risks and obstacles inherent in living alternatively, sets them apart from others in mainstream society. Surprisingly, to the outside observer, these communities function so well that they make their efforts seem easy; however, believing so would be misguided. Observers should keep in mind that these efforts at living alternatively are not easily undertaken; indeed, if they were, more people would be doing what they do.

In different ways and with varying success, each community has met challenges related but not limited to: (1) obtaining adequate financial capital and establishing sustainable economic systems; (2) cultivating relationships with outside social groups and gaining social legitimacy for their efforts, whether intended or not; (3) developing sustainable approaches to living in the face of degraded or insufficient natural resources; (4) addressing levels of consumption and waste production via methods of simple living and the use of appropriate technologies; and (5) integrating and employing collections of life skills and "learning-by-doing" competencies to build and make possible their community of choice. Consequently, by their example, a tremendously rich opportunity is afforded to people and communities in other bioregions who face the challenges of living more sustainably.

REFLECTIONS FOR OTHER COMMUNITIES

The work of these communities offers many examples for persons to begin to develop methods of their own in keeping with their specific bioregional needs. For example, as portrayed by the communities examined, others could consider the following:

- To consider the utility of using the McHarg Overlay Method (1971) toward examining the composition of their existing or developing community
- To reflect on their current lifestyle and consider how sustainable changes could be learned and integrated
- To question and critique the origin and quality of resources such as food and energy
- To think about the interrelationships between living and human economies in their bioregion
- Perhaps simply assess the limits of their knowledge and experience with regard to methods of sustainable living as an initial step toward taking greater responsibility for their life on earth

Furthermore, sustainable communities offer many examples of how social problems can be addressed and resolved. With these examples, other communities could learn from and employ the efforts of these communities in their own bioregions, including bioregionally mindful changes in diet and resource consumption, the development of sustainable economic enterprises, the affirmation of respectful and just relations with humans and non-humans alike, and the restoration of rich cultural histories into present-day living.[1]

All communities and their bioregions share a close and common bond with each other as they evolve and adapt to each other. Thus, use of bioregional "lenses" to define and examine communities can provide new insights about how communities function. Interestingly (as reflected indirectly in Arkin's 1995 work), when a community is examined from a bioregional context, the significance of its rural or urban qualification is transformed from simply being a descriptive term denoting location or demographics to that of a term describing the *relationship* between the community and its ecological base. Thus, if the bioregional context of communities were to be operationalized within community development efforts, the potential for creating healthy, symbiotic, and sustainable communities, such as those discussed herein, would likely grow.

Recent literature has established the relevance and salience of environmental quality issues to the causes of social justice and social welfare of communities (Hoff and McNutt, 1994). Thus, the linkages that exist between environmental issues and social welfare at local, regional, national, or international levels are of vital importance to community activists and community development practitioners, researchers, or educators who endeavor to be insightful and effective. Gaining a sophisticated understanding of and appreciation for these linkages may well prove to be vital in that, as time moves by and the current collage of social problems becomes further exacerbated by the pressures of acute environmental degradation, population growth, and resource depletion, the preparedness and competency of communities and nations and the global community's response will be given its ultimate test.

REFERENCES

Arkin, L. (1995). An eco-village retrofit for Los Angeles: Healing an inner-city neighborhood. In *Communities directory — A guide to cooperative living* (pp. 109–116). Langley, WA: Fellowship for Intentional Community.

Bane, P. (1996). The state of permaculture: A lot more than organic gardening. *The Permaculture Activist, 35* (November), 2.

Barnes, L. (1996). *1996 Katúah Bioregional Conversational Salon* (unpublished pamphlet available from the author at P.O. Box 1303, Waynesville, NC 28786).

Barnes, L. and Messick, R. (1991). The web of life — A Katúah almanac. *Katúah Journal,* Spring, 20–22.

Bates, A. (1988). Technological innovation in a rural intentional community, 1971–1987. *Bulletin of Science, Technology & Society, 8,* 183–199.

da Silva, A. (1996). From consensus to democracy: In for the long haul. *The Permaculture Activist, 35* (November), 6–8.

Eco-Village Training Center at The Farm (1996). *Examining sustainable design at The Farm intentional community* (Course flyer, 2).

Gaia Trust (1995). *What is an ecovillage?* Available at http://www.gaia.org/evis/whatisecovillage.html.

Gilman, R. (1995). Definition of an eco-village. In *Communities directory — A guide to cooperative living* (p. 109). Langley, WA: Fellowship for Intentional Community.

Global Eco-Village Network (1996). *The earth is our habitat — Proposal for support programme for eco-habitats as living examples of Agenda 21 Planning* (p. 6). Denmark: Gaia Trust.

Haenke, D. (1995). Bioregionalism and community: A call to action. In *Communities directory — A guide to cooperative living* (pp. 117–120). Langley, WA: Fellowship for Intentional Community.

Harte, J. (1993). *The green fuse — An ecological odyssey.* Berkeley: University of California Press.

Hoff, M.D. and McNutt, J.G. (Eds.) (1994). *The global environmental crisis: Implications for social welfare and social work.* Aldershot, England: Ashgate/Avebury Books.

Kahn, S. (1991). *A guide for grassroots leaders.* Silver Spring, MD: NASW Press.

Kleihauer, S. (1996). Walking life's Narrow Ridge. *Earthlight, 22* (Summer), 12.

Marsh, C. (1996). Planning for a new tribe. *The Permaculture Activist, 35* (November), 23–25, 28.

McHarg, I. (1971). *Design with nature.* Garden City, NY: Doubleday/Natural History Press.

Narrow Ridge Earth Literacy Center (1995). *Syllabus for the Narrow Ridge Center's "Earth Literacy Internship."* Washburn, TN: Author (Rural Route 2, Box 125, 37888).

Nickle, B. (1995). Narrow Ridge: A place to get back on. In M. Smith (Ed.), *Introducing Earth literacy — A letter to Al* (p. 37). Washburn, TN: Narrow Ridge Earth Literacy Center.

Nolt, J. (1995). *Down to earth: Toward a philosophy of nonviolent living.* Washburn, TN: Earth Knows Publications.

Nolt, J. (1996). *Understanding what we are doing to the upper Tennessee Valley and the southern Appalachian Mountains.* Paper presented at Centripedal Lecture, University of Tennessee, Knoxville, October 9.

Nolt, J., Bradley, A.L., Knapp, M., Lampard, D.E., and Scherch, J. (1997). *What have we done? The Foundation for Global Sustainability's state of the bioregion report for the upper Tennessee Valley and southern Appalachian Mountains.* Washburn, TN: Earth Knows Publications.

Ochre, G. (1996). Community glue. *Permaculture Magazine* (Clanfield, Hampshire, England: Permanent Publications), *12* (Summer), 12–14.

Smith, M. (Ed.) (1996a). *Introducing earth literacy — A letter to Al.* Washburn, TN: Narrow Ridge Earth Literacy Center.

Smith, M. (1996b). Planting new roots for education. *EarthLight, 22* (Summer), 10.

Sosadeeter, M. (1993). Designing with nature: The McHarg Overlay Method of Site Analysis. *The Permaculture Activist, 28* (February), 8–9.

Southern Appalachian Man and the Biosphere Cooperative (1996). Summary report. *Southern Appalachian Assessment.* July, Figure 1, 4.

Tatum, J.S. (1992). The home power movement and the assumptions of energy policy analysis. *Energy, 17*(2), 99.

Tennessee Valley Authority (1995). *Energy Vision 2020 integrated resource plan — Environmental impact plan,* Vol. 2. Chattanooga, TN: Author.

Traugot, M. (1994). Introduction. In *A short history of The Farm* (p. 1). Summertown, TN: The Farm.

Weil, M.O. (1995). Women, community, and organizing. In J.E. Tropman, J.L. Erlich, and J. Rothman (Eds.), *Tactics and techniques of community intervention* (3rd ed.), (pp. 118–134). Itasca, IL: F.E. Peacock.

Weil, M.O. (1996). Community building: Building community practice. *Social Work, 41*(5), 481–499.

Western North Carolina Alliance Old Growth Committee (1995). *Nantahala–Pisgah National Forests old growth survey: Citizen involvement in old growth protection.* Ashville, NC: Western North Carolina Alliance.

Wheeler, D. (1991). Economy/ecology. *Katúah Journal,* Spring, 1–3.

ENDNOTE

1. See Ochre's (1996) *Recipe for Community Glue* for a list of "ingredients" for consideration in developing sustainable communities.

Sustainable Urban Community Development Cases

CHAPTER 6

Toxic Risk, Community Resilience, and Social Justice in Chattanooga, Tennessee

Mary E. Rogge

The city of Chattanooga, in Hamilton County, Tennessee, has moved, in less than 30 years, from being one of the most polluted cities in the United States to receiving national and international prominence for its achievements in sustainable development. This chapter describes the mixture of organizational and individual efforts (private not-for-profit, for-profit, and governmental) that have dramatically reduced pollution in the air, land, and water of the Chattanooga–Hamilton County community. These efforts have positioned Chattanooga as an international model of an "environmental city" that is realizing sustainable development and its intricately connected components of "economy, ecology, equity" (*Sustainable Chattanooga*, 1997, p. 1).

Chattanooga's environmental metamorphosis attests to positive changes that can be achieved through concentrating community effort and resources on a problem. It also illustrates society's continuing struggle to attend both to the social welfare of citizens at large and to the particular needs for social justice for the most disenfranchised people. Chattanooga–Hamilton County, named as one of 21 exemplary U.S. communities through the President's Council on Sustainable Development (Glick, 1996), has within its boundaries Chattanooga Creek, one of the most highly polluted waterways in the southeast (Nolt et al., 1997). Consistent with research that shows disenfranchised populations to be disproportionately at risk from toxic waste dumps and other forms of pollution, citizens living nearest to the creek are predominantly low income and African-American. Community efforts to achieve equity and barriers to cleaning up the recalcitrant creek are chronicled here.

First, a brief history of Chattanooga–Hamilton County development with an emphasis on pollution and the distribution of toxic chemical releases and con-

1-57444-129-9/98/$0.00+$.50
© 1998 by CRC Press LLC

tamination within the area is provided. Second, community strategies and tactics used to clean up the community overall, and Chattanooga Creek specifically, are reviewed. Third, ways in which Chattanooga–Hamilton County has linked these strategies to ongoing action and future planning for sustainable development are described. Fourth, key successful strategies and barriers are summarized and discussed in the context of broader social phenomena that have contributed to the community's successes.

HAMILTON COUNTY CHARACTERISTICS

Hamilton County, Tennessee, is one of six counties in the Chattanooga, Tennessee–Georgia Bi-State Metropolitan Statistical Area (Natural Resources Defense Council, 1996). Located on the Tennessee–Georgia state line, the Chattanooga Metropolitan Statistical Area straddles the South–Deep South border, significant for historic distinctions in racial discrimination, political conservatism and economic structures (Dunn and Preston, 1991).

With 539 square miles and an estimated 1996 population of 296,183, the population density of this primarily metropolitan county is 550 people per square mile (U.S. Census, 1990/National Decision Systems, 1996). The population is heavily clustered in the southern part of the county, in and around the 15 to 20% of the county that is comprised of Chattanooga (Stockwell and Sorenson, 1994). About 73% of the community's citizens have graduated from high school and about 45% have some college-level education; the average age is 38, 19% are African-American and 79% are white, the median household income is $46,406, the unemployment rate is about 4.2%, and the ratio of white-to blue-collar workers is about 59 to 41 (U.S. Census, 1990/National Decision Systems, 1996).

A HISTORICAL PERSPECTIVE ON TOXIC RISK IN CHATTANOOGA–HAMILTON COUNTY

Chattanooga–Hamilton County is located in the 41,000-square-mile watershed of the Tennessee River. The nation's fifth largest river meanders through parts of the states of Kentucky, Tennessee, Mississippi, Alabama, Georgia, and North Carolina (Tennessee Valley Authority, 1995). Hamilton County terrain ranges from mountains and hills to flatter, fielded areas. Peopled by Cherokee and other Native American settlements from as early as 7,000 years ago, the area was noted for its beauty in DeSoto's exploratory annals in the mid-1500s (Armstrong, 1931; Livingood, 1981). Agricultural development was restricted in the area because of the terrain; however, the river and natural resources, such as coal, promoted fast development during the industrial era. Chattanooga, in the southern part of the county, was founded on the banks of the Tennessee in 1838. The city and surrounding area became a North/South industrial and transportation hub that was a focal control point during the Civil War. In the 1900s, the industrial nature of the area continued to expand, with an important boost from the Tennessee Valley Authority's (still a major employer) development of cheap hydroelectric power (Bass and DeVries, 1976). By 1974, Chattanooga was ranked nationally as eighth in industry per capita (Chattanooga–Hamilton County Air Pollution Control Bureau [C-HCAPCB], 1996).

Accompanying the city and county's high industrial ranking were extremely high national rankings in pollution levels. During the 1940s, 1950s, and 1960s, the pollution alone so limited visibility in Chattanooga that car headlights were often turned on midday. During the 1950s, women reported their nylon hose disintegrating when worn outdoors (Glick, 1996). The pollution was visible and increasingly linked to social welfare concerns, especially health. By 1963, the community's mortality rate from tuberculosis was three times the national average. Citizens' health concerns also included emphysema and other respiratory diseases (C-HCAPCB, 1996). Increased emissions from industrial and commercial processes and the use of soft coal for heating purposes were exacerbated by atmospheric temperature inversions that confined the pollution within the mountainous terrain surrounding the city.

In a 1969 U.S. Department of Health, Education and Welfare (DHEW) report (C-HCAPCB, 1996), Chattanooga was ranked as the most highly polluted city in the United States — first in air pollution from particulate matter releases (e.g., airborne smoke and dust; linked to respiratory and cardiopulmonary illness) and second only to Los Angeles, California, in ground-level ozone (toxic gas created by interactions of sunlight with nitrogen oxide and other chemicals; ozone is also linked to cardiopulmonary damage) (Harte et al., 1991).

A unique level of involvement across the community appears to have converged in the mid to late 1960s to turn local concern into a concerted series of actions. Citizens' groups, including the Hamilton County Tuberculosis Association and the North Hamilton County Air Pollution Committee, pressured local officials with registered complaints, held educational symposia to increase knowledge about health effects of pollution in the community, and held joint meetings — with media invited — in places where the pollution was easily visible. Chattanooga's high national pollution rankings were paired with increasingly available stories of illness and death from pollution in other cities, such as Donora, Pennsylvania (where a 1948 air inversion killed 20 citizens and caused illness in half the city's population). Around 1966, the efforts of the Chattanooga citizens' groups were significantly boosted by journalists associated with three Chattanooga newspapers. Newspaper reports visibly documented the pollution and helped to broaden city-wide interest and to "solidify public opinion in trying to correct the pollution problem" (C-HCAPCB, 1996, p. 11). Local officials then requested and received assistance from the U.S. DHEW to conduct two studies of pollution in the community. The second study resulted in the 1969 ranking of Chattanooga as the most polluted city in the country.

Citizens' efforts to strengthen local air pollution regulations were given a critical boost by the passage of the National Clean Air Act of 1965 and subsequent federal legislation, which mandated state plans in which localities participated. Among incentives provided by the federal legislation were prohibitions against allowing certain types of new industry into the community unless air quality standards were met and federal clean air funding, used locally to support the Chattanooga–Hamilton County air pollution control program.

Local industry's proactive commitment to achieving or bettering federal standards is credited as an essential part of the community's success in developing pollution reduction measures. In turn, a very important contributor to local industry's stance was the Chattanooga Chamber of Commerce's decision in the late 1960s to assert itself in the long-term development, enforcement, and marketing of air pollution standards and laws in the community (C-HCAPCB, 1996). Local industry, at a cost of approximately $40 million, almost uniformly met the standards by the 1972 deadline; by 1974, the community had been nationally recognized for its reductions in air pollution. Since 1989, Chattanooga–Hamilton County has met or exceeded federal primary health standards for the six nationally monitored pollutants (particulate, ozone, nitrogen dioxide, carbon monoxide, sulfur dioxide, and lead) considered to be both most dangerous to human health and nationally widespread (C-HCAPCB, 1993, 1994, 1995).

SUSTAINABLE VISION THROUGH PARTICIPATION AND PROCESS

Individual citizens and citizen groups, private ventures, and public entities have continued to concentrate on pollution reduction; these efforts are currently embedded in the *Sustainable Chattanooga* movement. The community's successful pollution reduction actions are widely recognized as the springboard to current sustainable development initiatives. In 1984, the community development initiative was reframed into an expansive community participation goal-building process, based in part on an Indianapolis, Indiana, initiative (Bernard and Young, 1997). *Chattanooga Venture,* a private, not-for-profit, thriving organization which is an amalgamation of citizens' groups, was created by community leaders for the express purpose of mobilizing, coordinating, and nurturing community action. *Chattanooga Venture* first initiated a community needs assessment and strategic planning process called *Vision 2000. Vision 2000* gathered information from over 1,000 citizens, summarized their thoughts into 40 future goals, and, through *Chattanooga Venture* leadership, organized new and existing organizations and leaders into a series of citizen task forces and organizations to address the goals. Among the outcomes of this process have been the Tennessee Riverpark and Aquarium, locally built zero-emission electric buses that are used locally and are increasingly sold nationally and internationally, and Chattanooga-hosted conferences that have brought expert environmental consultants to the community. The *Vision 2000* process spawned the community's first domestic violence shelter, the nationally recognized low- and moderate-income housing initiative of Chattanooga Neighborhood Enterprise, and the Neighborhood Network support system for neighborhood groups (Sustainable Development Online: Chattanooga Community Link [SDO], 1996a–c; Bernard and Young, 1997; Vaughan, 1996).

In 1993, *Revision 2000* gathered new information from a representative sample of 2,600 citizens and derived 27 new goals, 122 recommendations, and the broadly supported, citizen-based vision of being a nationally recognized "environmental center" (SDO, 1996a, p. 1). In 1996, the broad-based community planning and development process was advanced with *FUTURESCAPE 1996,* another stage of community planning implemented through the Chattanooga–Hamilton County Regional Planning Commission and supported by the Chamber of Commerce and *Chattanooga Venture.* Through a survey of over 3,000 citizens and public forum techniques, *FUTURESCAPE* assessed community thinking about ways to change local governmental ordinances and codes to "encourage the kind of development we want and to discourage the kind of development we don't want" (Vaughan, 1996, p. 5).

Measurable advances and widespread, multifaceted community participation have become a reality through this ongoing community visioning process. *Chattanooga Venture* is the strategic core link in a web of organizational and individual efforts that are synergistically advancing the quality of living in Chattanooga–Hamilton County.[1] In addition to the outcomes described above, the visioning process is producing:

- The Orange Grove Recycling Facility, which employs over 100 individuals with disabilities and provides education tours regarding disabled individuals' social contributions and illustrates environmental problems
- A new prototype for a city–county school system
- Environment and children's health-enhancing recreational alternatives, especially for children in low-income neighborhoods
- Plans for the redevelopment of chemically contaminated brownfields with eco-industrial parks
- Preservation of natural resources, including the Tennessee River Gorge and rejection of ecologically destructive enterprises such as chip mills
- Neighborhood development, including the inner city Westside Community Revitalization

Through a 35-member development corporation, 7 of whom are neighborhood residents, the $21 million revitalization has to date renovated a closed school into a healthcare facility and established a community center with a business development program.

Chattanooga–Hamilton County has benefited from diversification and an ecological systems perspective. For example, rather than the selection of a few priority goals toward which to focus community resources, the visioning process encourages the derivation from community members of a much fuller set of goals and projects and then actively supports the creation of entities to implement them. Funding for projects, programs, and conferences is garnered from an impressive array of city, county, state, and national and international governmental entities, businesses, local organizations and foundations, including the Junior League and United Way, and national foundations, including the Robert Wood Johnson Foundation. Tactics and strategies, customized to match projects, range from international business marketing strategies for locally produced electric buses to the formation of "Household Eco-Teams" of six to eight households and a coach who support each other in improving individual and household sustainable living (SDO, 1996a; Nolt et al., 1997, p. 231).[2]

The *Sustainable Chattanooga* movement has been visibly empowering to many citizens. Chattanooga–Hamilton County citizens are active locally, statewide, nationally, and internationally in the promotion of improved quality of life. For example, in 1996, the community hosted a combined national and international Environmental Summit that included representation from United Nations participants and a national conference of zero-emissions industries. Also in 1996, leaders of the Neighborhood Network served as representatives to a United Nations international conference on the environment in Turkey (G. Spring, personal communication, June 1996). Milton Jackson, leader of Stop Toxic Pollution (S.T.O.P.), serves as one of three community consultants to the national Board of Scientific Counselors for the U.S. Agency for Toxic Substances and Disease Registry (ATSDR, 1996). In acknowledgment of the economic, environmental, and social synergy that has occurred in this community, Chattanooga–Hamilton County has been named one of 21 exemplary U.S. communities under the President's Council on Sustainable Development (PCSD) (Glick, 1996).

Chattanooga–Hamilton County's far-reaching citizen interest and action are consistent with what many citizens consider to be a long-standing local culture of people moving past traditional boundaries to work together (G. Spring, personal communication, 1996). Many, however, trace the birth of the current strong network of sustainable community development initiatives to the murky, harsh 1960s when Chattanooga was the most polluted city in the United States — when citizens, media, and federal legislation pressured and joined with local business, industry, and government leaders to clean up the city's environmental act. The unique success of Chattanooga's sustainable development ventures owes a debt to that time when "success in cleaning up our air gave us the confidence to believe we could succeed in other ways" (Vaughan, 1996, p. 1).

TOXIC RISK, SOCIAL WELFARE, AND JUSTICE:
THE SLOW LANE INTO THE SUSTAINABLE FUTURE?

Chattanooga–Hamilton County has made tremendous progress in improving the social welfare of its citizens through reducing pollution. The community continues, however, to struggle to achieve social and environmental justice for African American citizens.

The social welfare benefits of reducing pollution are made vividly — visibly — clear with images of Chattanooga at its polluted worst: car headlights on at noon, women's nylons disintegrating, high rates of respiratory illness. Although the Chattanooga–Hamilton County community continues efforts to reduce pollution, there remain health risks from toxic emissions. In 1992, more than 5 billion pounds of toxic chemicals was released into the air, on land, or in water from commercial and industrial facilities (Toxic Release Inventory, 1992). Chattanooga–Hamilton County has yet to remain below federal secondary standards for particulate matter releases associated with respiratory and cardiopulmonary illnesses (Tennessee Valley Authority, 1995). Particulate matter releases in Chattanooga–Hamilton County are still the highest of the five metropolitan statistical areas in Tennessee and are estimated to cause 63 cardiopulmonary deaths annually (Natural Resources Defense Council, 1996).

The adequacy of social justice is at issue in Chattanooga–Hamilton County. The main sources of toxic chemical emissions are clustered in the same part of town as its African-American citizens. Thirty-one of the 45 commercial and industrial facilities emitting toxic chemicals in Chattanooga–Hamilton County are clustered in Chattanooga's urban downtown area. Three of these facilities ranked among the top 50 emitters in the eight southeastern states of chemicals considered to cause developmental and neurological damage (Eckert and Stockwell, 1993). These facilities cluster in census tracts where the community's African-American population is also concentrated (Stockwell and Sorenson, 1994). Within this area is also Chattanooga Creek, the most polluted area in Chattanooga–Hamilton County and the most polluted waterway in the southeastern United States (SDO, 1996a; Nolt et al., 1997). Currently, of the 5,331 residents of the two residential census tracts, Piney Woods and Alton Park, that border the creek, about 98% are African-American. A health center, a high school, and a middle school are located directly alongside the creek.

Chattanooga Creek, a major tributary of the Tennessee River, was the site for most of the community's industries for about 70 years. The facilities along Chattanooga Creek, including "coke manufacturers, organic chemicals, wood preserving, metallurgical and foundry operations, tanning and leather products, textiles, brick making and pharmaceuticals" facilities, released chemicals into the air, on the adjacent land, and in Chattanooga Creek. The Department of Defense dumped coal tar deposits into Chattanooga Creek during World War II, and the creek's floodplain was widely used as a waste dump by municipalities, industries, and individual citizens (SDO, 1996a). In September of 1994, a 7.5-mile stretch of the creek between the Tennessee–Georgia border and the Tennessee River was targeted for hazardous waste cleanup as a U.S. Superfund National Priority List site (U.S. Environmental Protection Agency [EPA], 1996). There are 12 other creek sites designated as state Superfund sites and 29 more sites under consideration for designation (Tinker et al., 1996).

During the 1950s, neighborhoods around Chattanooga Creek began to experience white flight and immigration of a primarily African-American population. Chattanooga–Hamilton County citizens from both within and outside the Piney Woods and Alton Park neighborhoods report that race, income, and toxic risk have been intricately connected in over 30 years of concern, frustration, and battles to improve living conditions in these neighborhoods. A series of Piney Woods and Alton Park citizens' groups (e.g., Alton Park/Piney Woods Community Coalition, Turning Point, Progressive Improvement League of South Chattanooga) fought toxic risk problems in the 1970s and 1980s and played important roles in advocating with local, state, and federal entities for a series of health surveys done in the neighborhoods. The findings of the surveys, performed by a mix of privately contracted, state, and federal entities, ranged from no significant health threats from outdoor air releases to significantly elevated self-reported health problems to significant dangerous indoor releases of nitrogen dioxide within neighborhood public housing units, caused by "malfunction or misuse of gas appliances" (C-HCAPCB, 1996, p. 27). Other citizen groups with state and national connections have been instrumental in bringing resources to the residents near the creek. For example, Dr. Grace Hewell of Chattanooga used her Washington, D.C.–based Health Policy Group to advocate with federal officials for a study of the area. As a result, the EPA completed a 1994 EPA Region IV study of toxic risk in Chattanooga (Stockwell and Sorenson, 1994). The effort to initiate the EPA Region IV study also involved the National Council of Negro Women and Greenpeace USA (G. Hewell, personal communication, 1995).

In 1994–95, the Piney Woods/Alton Park neighborhood S.T.O.P. group successfully advocated for additional federal agency health assessments in the neighborhoods because of concerns about "cancer, miscarriages, birth defects, breathing problems, headaches, eye and skin irritations" (ATSDR, 1995, p. 10). Of particular concern were local children swimming in the creek, neighbors and non-locals fishing in it, and homeless men camping in wooded areas along the creek. Homeless men have been treated in local clinics for external and internal chemical burns associated with bathing in the creek and drinking the water (Johnson, 1995; Tinker et al., 1995). ATSDR chemical analyses found "high levels of polyaromatic hydrocarbons, pesticides, metals and polychlorinated biphenyls" in creek water and sediments, in fish, and in the air near the creek (ATSDR, 1995, p. 10).

The ATSDR, in conjunction with neighborhood contacts through S.T.O.P., the schools, and other neighborhood groups including the Bethlehem Center, performed a community needs assessment and carried out an extensive series of public education activities for neighborhood schools (students, teachers, and nurses), the area's homeless shelter, the Alton Park Health Center, and other local health facilities. Activities included a series of science lessons for teachers, contracting with neighborhood schools and teachers to teach yearly about the toxic hazards of the creek, and a "Don't Be a Creek Geek" poster contest. About 1,200 elementary school children received education about the creek through classroom activities and school assemblies. With the help of the schools, public libraries, media, and community organizations including S.T.O.P., the Bethlehem Center, and Community for a Cleaner Creek, over 3,000 fact sheets and maps showing the location of schools and homes relative to the creek were distributed to children and families. New signs advising people about the health hazards of the creek replaced old ones along the creek, and community meetings to discuss the creek were publicized through the media. S.T.O.P. played a key role as an outreach, communications, planning, and organizing liaison between ATSDR and neighborhood adults, teenagers, and children (SDO, 1996a; Tinker et al., 1996). Bethlehem Center, in addition to distributing the fact sheets, provided meeting space for public organizing meetings about the creek. The Bethlehem Center continues such activities in support of cleaning up the creek, with funds from a $20,000 grant from the Women's Division of United Methodist Church for environmental and health projects (Bethlehem Center–Chattanooga, 1997).

The focus of organizing activities in the Chattanooga Creek neighborhoods, as illustrated in the preceding paragraphs, was initially on assessing the extent and nature of toxic risk from Chattanooga Creek, then shifting to an emphasis on community education and behavioral change (i.e., "stay away from the creek"). As more information was more widely spread throughout the neighborhoods and among community activists and leaders, such as Milton Jackson of S.T.O.P. and Deborah Matthews of the Alton Park/Piney Woods Neighborhood Improvement Corporation, the focus shifted to neighborhood demands for action to clean up the creek and to improve neighborhood economic conditions. In August 1996, for example, the Neighborhood Improvement Corporation enacted a "demonstration of solidarity," demanding an environmental impact study of the adequacy of the proposed mitigation strategy for a state Superfund site which would put children of two of the neighborhood schools and a neighborhood daycare center at risk. Groups participating in the demonstration, which included the Chattanooga-based Southeast Center of Ecological Awareness, the NAACP, Operation PUSH, and Greenpeace, claimed environmental racism as an issue in the situation (Spear, 1996, p. A12).

Chattanooga Creek neighborhood groups continue organizing and development efforts to address the contamination of the creek and its surrounding neighborhoods, in varying degrees of involvement and success. The range of actors with whom they work includes the larger community-based grass-roots clearinghouse Neighborhood Network; local, state, and federal legislators and officials including representatives of the Tennessee Department of Public Health, the regional EPA office, and the U.S. EPA Office of Environmental Equity; and regional colleges and universities. Milton Jackson, a key figure in S.T.O.P. and in getting ATSDR intervention in the Chattanooga Creek neighborhoods, was appointed as a community consultant to the national ATSDR Board of Scientific Counselors, where his role includes strengthening partnerships between the ATSDR and communities across the United States as well as serving as liaison for ongoing ATSDR intervention with Chattanooga Creek.

The link between race, economic resources, and toxic risk in the Piney Woods/Alton Park neighborhood has been a sensitive, volatile issue, according to citizens who were interviewed for this study. Racism and economic oppression are important historic and current factors in community efforts to address toxic risk in the neighborhood. Chattanooga citizens living "inside" Piney Woods/ Alton Park and those living "outside" this neighborhood discuss these factors in terms of trust, control, and participation. Both "insiders" and "outsiders" describe insiders' distrust and skepticism of outsiders' involvement to ameliorate the toxic risk. In the initial approval process of a recent Tennessee Department of Public Health survey of the creek, for example, a critical issue for the insiders was proof that the planned activities were in no way a current-day version of the insidious Tuskegee experiments. Trust and control are inextricably related. In 1995, conflict erupted between insiders and outside researchers from a local college over the use of a U.S. EPA environmental equity grant. Insiders demanded the funds be used to pay neighborhood citizens; to create neighborhood jobs, business, and community relations efforts; and to clean up the contamination, rather than to pay researchers to study and educate the neighborhood yet again. A neighborhood leader stated, "We no longer need anybody that doesn't eat, sleep and *live* in this community to come in with their pity and tell us what *our* community needs" (Henderson, 1995, p. E1).

Both insiders and outsiders have described the neighborhood as isolated. Insiders may often feel their voices have not be heard, and outsiders may often perceive the insiders' lack of trust and hope to be a significant barrier to sustained participation in change efforts (SDO, 1996a). These issues of trust, control, skepticism, and participation described by community members have not, however, been endemic to all insider–outsider joint efforts. The S.T.O.P./ ATSDR collaboration, for example, was considered a "very positive partnership" (ATSDR, 1995, p. 11).

The social and political tensions associated with social justice, environmental racism, and community development in the Chattanooga–Hamilton County community extend beyond the boundaries of Piney Woods and Alton Park, and most likely beyond the southeastern region of the United States. For example, the 1994 EPA Region IV report on Chattanooga concluded that minority status and toxic risk are related (Stockwell and Sorenson, 1994). This report and a companion piece on Mobile, Alabama, proved to be highly controversial. Although the reports are public documents, a complex set of political tensions resulted in at least one EPA official threatened with loss of job, a gag order on discussions of the data, and suspension of further data collection on race, income, and toxic risk. On the other hand, Chattanooga–Hamilton County has exposed its environmental successes — and unresolved problems such as the toxic risk of significant portions of the community's African-American population around Chattanooga Creek — to international scrutiny.

The contamination of Chattanooga Creek is a dauntingly complex "technical, economic, and social" challenge (SDO, 1996b, p. 2). The science and equipment needed to adequately clean up 70 years of combined industrial, commercial, and citizens' wastes is still developing (Johnson, 1995). The technology of toxic cleanup raises interesting comparisons with the technology of toxic source reduction and prevention, such as pollution reduction measures in Chattanooga–Hamilton County. Economically, the costs of cleaning up Chattanooga Creek must factor in the costs and benefits to Piney Woods and Alton Park neighbors, other Chattanooga–Hamilton County citizens, businesses, and a wide range of other actors, in the context of the community's commitment to sustainable change. Socially, the recalcitrant issues of racial discrimination, distribution of economic resources, and other aspects of social vulnerability must be addressed to realize the community's vision of equitable, sustainable development.

CATALYST FOR SUSTAINABLE DEVELOPMENT: UNEVEN RATES OF CHANGE

The concentration of high social vulnerability and toxic risk in the Chattanooga Creek area strongly suggests that the social welfare benefits associated with the community's drive toward "environment, economy and equity" is unevenly distributed and leaves some of its most socially vulnerable citizens lagging behind in the move toward sustainable change. There is, however, continuing effort to remediate chemical contamination in the Chattanooga Creek area. Local and state officials, the U.S. EPA, and the ATSDR have cooperated in the establishment of a neighborhood-based repository of data and other environmental materials in the Alton Park Health Center. Industrial neighbors to these socially vulnerable citizens have met primary air standards, improved hazardous waste discharges, and paired with Piney Woods/Alton Park citizens in a neighborhood park project (SDO, 1996a). As noted earlier, activity surrounding the Piney Woods/Alton Park/Chattanooga Creek neighborhoods appears to be moving slowly from stages of assessment to public awareness and education to remedial and mitigatory action. Current efforts center around the complicated, time-consuming, expensive Superfund site remediation process (U.S. EPA, 1996). Citizens of the neighborhoods, however, do well to retain skepticism: new controversy developed in late 1996 when the Chattanooga Coke Plant site was removed from the scheduled U.S. Superfund National Priority List.[3]

As part of the greater Chattanooga–Hamilton County visioning process, plans for the Chattanooga Creek area include education, job training, and financing for small business development initiatives through banking/neighborhood partnerships, the U.S. Department of Housing and Urban Development, the Chattanooga Office of Economic and Community Development, the Chattanooga–Hamilton County Regional Planning Commission, and the Chattanooga–Hamilton County Community Enterprise program (Glick, 1996). Overall, however, the Chattanooga–Hamilton County sustainable development successes raise the question as to what positive change might have been instituted in the Piney Woods/Alton Park/Chattanooga Creek neighborhoods had the same high level of commitment, resources, and sense of urgency that created the community-wide air quality turnabout been focused there.

CONCLUSIONS

Generally, *Sustainable Chattanooga* enterprises currently rely on collaborative strategies and tactics typically associated with locality development and social planning methods of community organization (Rothman, 1995). In contrast, the Chattanooga Creek neighborhoods, with a history of economic and racial disenfranchisement, are more reliant on confrontative social action strategies to make their voices heard. Other essential principles of community organizing are reflected in the Chattanooga Creek neighborhoods' experience with the ATSDR intervention: the necessity of intensive collaboration and joint decision making with community members from the outset, the need for community investment in proposed interventions, and the importance of evaluating the outcomes of the intervention (ATSDR, 1995).

The organizing concepts at the core of Chattanooga–Hamilton County's sustainability successes are encapsulated in a President's Council on Sustainable Development discussion of lessons learned about community participation. The "lessons learned" include the importance of having community-wide participation, coordinating interrelated systems, maintaining a balance of vision and action, nurturing public/private collaboration, and finding positive working solutions (SDO, 1996c, p. 2). These concepts are elaborated below to summarize the activities described in this chapter.

Citizen Participation

First, community-based development demands extensive representative citizen participation from the outset. Chattanooga–Hamilton County expanded and nurtured community-wide participation through the non-governmental, core community organization of *Chattanooga Venture* and its community visioning and action processes. The expectation of involvement has been lodged in the general community mind-set through information-gathering techniques including an ongoing series of public forums where people meet face to face and through action plans that acknowledge and mobilize existing community groups and involve citizens in creating new ones. The lack of citizen involvement can cause serious conflict, mistrust, and interventions that are a poor fit for community needs and wishes. For example, the anger of the Piney Woods/Alton Park neighbors regarding the allocation of the U.S. EPA environmental equity grant, described earlier, arose in part because the neighbors had not known the grant had been submitted (Henderson, 1995). Overall, the Chattanooga–Hamilton County process enhances accountability, ownership, empowerment, and continued involvement. Chattanooga–Hamilton County has chosen to struggle with the issues of power and control that are inherent in achieving genuine, broad-based community participation. That the process is not always smooth is exemplified in a statement about the struggle to improve the community's air quality: "it is possible for public opinion to be very disruptive to the process of progress" (SDO, 1996a, p. 2).

A Vision of Interactive Systems

Second, Chattanooga–Hamilton County has created a clear vision of sustainability that is broadly and flexibly defined. Environment, economy, and social equity are conceptualized as three interlockings systems through which "to seek opportunities in its entire social fabric" (SDO, 1996c, p. 3). This has allowed citizens who have helped develop programs as different as a domestic violence shelter, international markets for electric vehicles, and the preservation of a river gorge to appropriately share the title of citizens working toward a sustainable future. Early decisions in the 1960s to embrace the compatibility of the environmental and economic components of sustainability required calculated risks and tremendous investments by local industry — a commitment shared by few other communities — which gave Chattanooga–Hamilton County a jump-start toward sustainability. The clear, flexible vision of the community has enriched its diversity in goals, resource development, participatory methods, and tactics for change. One of the potential pitfalls of such a flexible vision, however, is a diffusion of focus to the point that sustainability becomes a catchall concept. *Chattanooga Venture* and the periodic visioning processes serve as gatekeepers to keep initiatives from straying too far from the basic components of "environment, economy, equity."

Balance of Vision and Action

Third, the community has developed processes through which to balance its long-term vision with current action. Compared to the general tendency in U.S. society, in which current demands and gains typically eclipse future needs, the degree of widespread citizen strategic planning and visioning in Chattanooga–Hamilton County is astounding. If the community had not acted successfully upon many of the projects selected through its visioning, citizens' confidence in the long-term planning process would likely have been lost, and the critical balance of action and vision destroyed. The strategy of exchanging ideas on national and international levels is another useful technique for stimulating new thinking about both immediate and long-term problems and opportunities.

Coalition Building

Fourth, coalition building has been successful across many internal community boundaries, such as those between public and private organizations, private and not-for-profit enterprises, business entrepreneurs and environmentalists, and, although perhaps to a lesser degree, racial and economic boundaries. An additional benefit of such a community environment in which organizational collaboration is high is the ability to diagnose areas for intervention, such as neighborhoods or interest groups, such as the Chattanooga Creek neighborhoods, which may be more isolated from community activity and resources.

This community's links with external resources are vast, with regional, state, national, and international connections in business, government, and environmental organizations. The importance of state and national advocacy links, such as Chattanooga Creek advocates' connections with state legislators and federal environmental agencies, to the community's progress cannot be underestimated as a crucial strategy for community change.

Optimism About Locating Solutions

Another perspective integral to Chattanooga–Hamilton County's progress has been a positive attitude that there is a solution for every problem. This attitude is a futuristic, sustainable version of the "can do" American attitude. From its inception in 1984, the community's visioning process has regularly sponsored exchanges with experts in various environmental and economic areas. Public forums, local hosting of national and international conferences, and community representation at other conferences have been part of the means of locating existing solutions to local problems. The community has skillfully cross-fertilized learning opportunities and new perspectives on issues by *hearing* citizen voices in the visioning process; *establishing* an extensive, diverse web of citizen task forces and organizations working to improve various community conditions; and *coordinating* an interrelated systems approach.

Contextual Contributors

Finally, several social phenomena occurring across the United States and the world helped set the foundation for change in this community: federal regulations, the social movement against environmental racism, and the combined effects of information technology and globalization. Federal legislation has been a crucial tool for community organizing and development. For example, the Clean Air Act, which created incentives and promised penalties for local non-compliance, directly contributed to the community's drastic air improvement and to the groundwork for the sustainable development initiatives. The relatively young environmental racism movement, key events of which took place in the southeast region of the United States (U.S. General Accounting Office, 1983; Bullard, 1990), has provided Chattanooga Creek activists with a new "name" for the inequitable conditions in their neighborhoods and with support from organizations such as Greenpeace USA. Additionally, federal mandates regarding environmental equity and community partnerships have been levers used successfully by the Chattanooga Creek activists to support their efforts. Finally, *Sustainable Chattanooga* has embraced many opportunities made available through advances in information technology and globalization. The community continues to seek out new technologies and has created a sustainable niche by courting environmentally proactive businesses, marketing its electric buses internationally, and planning for a zero-emissions eco-industrial park. Additionally, the community and many of its citizens and organizations are making extensive use of the Internet for the exchange of ideas and marketing.

Chattanooga–Hamilton County has made tremendous progress through amalgamated, if at times uneven, efforts from all sectors of the community. Its selected developmental strategies have made the concept of sustainability and its triumvirate components of environment, economy, and equity virtually household words in the community. A complex web of individual and organizational efforts has been created, each strand of which, in its diversity, is building toward the common goal of sustainable development. Of course, there is much to do, including remediation of the problems facing the Chattanooga Creek neighborhoods. Further, despite growing population pressures in the southeast and worldwide, even *Sustainable Chattanooga* does not emphasize population management and reduction in its visionary planning. This is a community, however, that will continue to embrace new concepts and challenges as it continues its course to further develop environment, economy, and equity in all its sectors.

AUTHOR'S NOTE

The information reported here was gathered as part of the author's doctoral dissertation at the George Warren Brown School of Social Work, Washington University, St. Louis, Missouri. The study used structural equation modeling to examine the relationship of toxic chemical releases from industrial and commercial facilities with social vulnerability indicators in the eight southeastern states. The qualitative case study reported in this chapter is based on interviews and documents gathered primarily in 1996 and updated in 1997. I am grateful to interviewees, dissertation chair Dr. David F. Gillespie, committee members, and others who contributed their valuable time, thoughts, and perspectives to this work.

REFERENCES

Agency for Toxic Substances and Disease Registry, Board of Scientific Counselors (1995). Minutes. *ATSDR science corner.* April. Available at http://atsdr1.atsdr.cdc.gov:8080/1995am.html.

Agency for Toxic Substances and Disease Registry, Board of Scientific Counselors (1996). Minutes. *ATSDR science corner.* November. Available at http://atsdr1.atsdr.cdc.gov:8080/1996am.html.

Armstrong, Z. (1931). *The history of Hamilton County and Chattanooga, Tennessee.* Chattanooga, TN: Lookout Publishing.

Bass, J. and DeVries, W. (1976). *The transformation of southern politics: Social change and political consequence since 1945.* New York: Basic Books.

Bernard, T. and Young, J. (1997). *The ecology of hope.* Gabriola Island, BC, Canada: New Society Publishers.

Bethlehem Center–Chattanooga (1997). *Sallie Crenshaw Bethlehem Center: United Methodist Neighborhood Center.* Available at http://www.chattanooga.net/bethcent/index.html.

Bullard, R.D. (1990). *Dumping in Dixie: Race, class, and environmental quality.* Boulder, CO: Westview Press.

Chattanooga–Hamilton County Air Pollution Control Bureau (1993). *1993 progress report.* Chattanooga, TN: Author.

Chattanooga–Hamilton County Air Pollution Control Bureau (1994). *Progress report 1994.* Chattanooga, TN: Author.

Chattanooga–Hamilton County Air Pollution Control Bureau (1995). *Progress report 1995.* Chattanooga, TN: Author.

Chattanooga–Hamilton County Air Pollution Control Bureau (1996). *A history of air pollution control in Chattanooga and Hamilton County.* Chattanooga, TN: Author.

Dunn, J.P. and Preston, H.L. (Eds.) (1991). *The future south: A historical perspective for the twenty-first century.* Urbana: University of Illinois Press.

Eckert, J.W., Jr. and Stockwell, J.R. (1993). Categories of released chemicals reported to the Toxic Release Inventory: 1990 data. Emergency Planning and Community Right-to-Know Act. U.S. Environmental Protection Agency Region IV. Atlanta, GA: U.S. Environmental Protection Agency Region IV.

Glick, D. (1996). Cinderella story. *National Wildlife, 34*(2), 42–46.

Harte, J., Holdren, C., Schneider, R., and Shirley, C. (1991). *Toxics A to Z: A guide to everyday pollution hazards.* Berkeley: University of California Press.

Henderson, V. (1995). Alton Park residents want active cleanup role. *Chattanooga Free Press,* November 2, p. E1.

Johnson, B.L. (1995). ATSDR: Public health actions and findings. Congressional testimony by Assistant Surgeon General, Assistant Administrator, Agency for Toxic Substances and Disease Registry, Public Health Service, U.S. Department of Health and Human Services, June 27. Available at http://atsdr1.atsdr.cdc.gov:8080/test627.html.

Livingood, J.W. (1981). *Hamilton County.* Tennessee County History Series. Memphis, TN: Memphis State University Press.

Natural Resources Defense Council (1996). *Breath-taking: Premature mortality due to particulate air pollution in 239 American cities.* May. Available at http://www.nrdc.org/trbreath/state/TN.html.

Nolt, J., Bradley, A.L., Knapp, M., Lampard, D.E., and Scherch, J. (1997). *What have we done? The Foundation for Global Sustainability's state of the bioregion report for the Upper Tennessee Valley and the Southern Appalachian Mountains.* Washburn, TN: Earth Knows Publications.

Rothman, J. (1995). Approaches to community intervention. In J. Rothman, J.L. Erlich, and J.E. Tropman (Eds.), *Strategies of community intervention* (pp. 26–63). Itasca, IL: F.E. Peacock Publishers.

Sohn, P. (1996). Creek Superfund site cut in half: Court takes Mead plant grounds off cleanup list. *The Chattanooga Times,* November 15, pp. B1, B2.

Spear, K.L. (1996). Meeting will protest pollution at Howard School. *Chattanooga Free Press,* August 18, p. A12.

Stockwell, J.R. and Sorenson, J.W. (1994). *The Chattanooga, TN/GA metropolitan statistical area (MSA) toxic release inventory — Geographic information system (TRI-GIS) comparative risk screening analysis.* U.S. Environmental Protection Agency, Region IV Report. Atlanta, GA: U.S. Environmental Protection Agency.

Sustainable Chattanooga (1997). Available at http:/bertha.chattanooga.net:80/SUSTAIN/index.html.

Sustainable Development Online: Chattanooga Community Link (1996a). *President's Council on Sustainable Development briefing book.* Available at http://www.chattanooga.net/SUSTAIN/pcsd_briefing_book /html.

Sustainable Development Online: Chattanooga Community Link (1996b). *President's Council on Sustainable Development briefing book.* Available at http://www.chattanooga.net/SUSTAIN/pcsd_briefing_book /resources_summary.html.

Sustainable Development Online: Chattanooga Community Link (1996c). *President's Council on Sustainable Development briefing book.* Available at http://www.chattanooga.net/SUSTAIN/pcsd_briefing_book /particihousing_summary.html.

Tennessee Valley Authority (1995). *Energy Vision 2020: Integrated resource plan environmental impact statement.* Chattanooga, TN: Author.

Tinker, T., Lewis-Younger, C., Isaacs, S., Neufer, L., and Blair, C. (1995). Environmental health risk communication: A case study of the Chattanooga Creek site. *Journal of the Tennessee Medical Association, 88*(9), 343–349.

Tinker, T., Moore, G., and Lewis-Younger, C. (1996). Chattanooga Creek: Building partnerships for a healthier environment. *The Tennessee Conservationist, 1,* 14–18.

Toxic Release Inventory CD-ROM (1992). *Toxic Release Inventory CD-ROM technical documentation.* Washington, DC: U.S. Environmental Protection Agency.

U.S. Census, 1990 /National Decision Systems (1996). *In Chattanooga & Hamilton County Tennessee: Facts at a glance.* Available at http://www.rivervallypartners.com/facts.html.

U.S. Environmental Protection Agency (1996). *EPA National Priority List.* Available at http://www.epa.gov/superfund/oerr/impm/products/nplsites/html/ 0403765n.html.

U.S. General Accounting Office (1983). *Siting of hazardous waste landfills and their correlation with the racial and socio-economic states of surrounding communities.* Report No. GAO/RCED-83-168. Washington, DC: U.S. Government Printing Office, June.

Vaughan, J.G., Jr. (1996). *Sustainable Chattanooga: Defining a new model for community success.* Presentation by Chattanooga Area Chamber of Commerce President to American Chamber of Commerce Executives, Norfolk, VA, October 6.

ENDNOTES

1. See, for example, the extensive development of the city's *Sustainable Chattanooga* Internet site at http://bertha.chattanooga.net:80/SUSTAIN/index.html. The site's sections regarding the President's Council on Sustainability are particularly informative in

understanding the structure, organizations, development, and products derived through the community visioning process

2. "Household Eco-Teams" are the change agent mechanisms of the Global Action Plan, being developed in 13 countries by the United Nations Environmental Programme. In the United States, the U.S. Global Action Plan for the Earth and the President's Council on Sustainable Development (PCSD) are co-sponsors of the plan. Chattanooga was selected by the PCSD to be the first southeastern region community in the United States. Funding for the plan in Chattanooga is currently through the Chamber Foundation and the Chattanooga Public Works Department; additional funds are being solicited from a range of public and private sources (Nolt et al., 1997).

3. Twenty-three acres of the Chattanooga Creek U.S. Superfund site, where the Department of Defense produced coal tar that was dumped into the creek, is now owned by the Mead Corporation. The U.S. Court of Appeals, in its decision to remove the plant site from the Superfund site, noted that the Mead Corporation is not sufficiently linked to the plant's contamination of the creek to hold it liable. The EPA may appeal or file for separate Superfund site status for the 23 acres (Sohn, 1996).

CHAPTER 7

The Contribution of Universities to Building Sustainable Communities: The Community University Partnership

Michelle Livermore and James Midgley

Today, poverty, deprivation, physical decay, economic isolation, and numerous other social ills plague urban communities throughout the world. As government social programs in many industrial countries have been retrenched, inner-city poverty has become more severe. For this reason, attempts are again being made to mobilize community resources for sustainable local development. Community development is again being promoted as a means for addressing the pressing problems facing urban communities.

Community practice has a long history. Social workers and others have been engaged in community organizing for many decades. However, it is only in relatively recent times that the need for local economic development has been fully recognized. While much community practice has in the past focused on issues of mobilization, organizing, and participation, it is now more widely accepted that activities designed to enhance people's participation in decision making need to be accompanied by programs that improve their material well-being. It is also now more widely recognized that community social programs need to be directly linked to economic development efforts and environmental restoration and protection, and that development at the local level needs to have a long-term, sustainable commitment. In the past, community interventions often involved the provision of short-term projects, funded through external resources. When these resources were depleted, the projects were often terminated. Today, sustainability is a critical aspect of local community intervention. It is now widely recognized that local development effort must not only involve local people in decision making and implementation, but must ensure that these efforts become viable in the long run.

1-57444-129-9/98/$0.00+$.50
© 1998 by CRC Press LLC

The need for local, sustainable environmental, economic, and social development is now being promoted by many different organizations. A recent innovation is the involvement of universities in local community development. Although universities were previously engaged in community-level activities, particularly through the settlement movement, there has been a renewed interest in using university resources to assist local communities to engage in sustainable development.

The universities which have been in the forefront of community outreach are mostly located in the urban areas of large cities marked by a high degree of poverty and deprivation. Acutely aware of the problems associated with urban decay, these universities have embarked on various activities designed to assist their neighboring communities. One example is the University of Pennsylvania, which has been engaged in a major community outreach project focused on local schools (Harkavy and Pucket, 1994). Many other urban universities have since established similar projects. Universities situated in small college towns have followed this example by establishing outreach projects in low-income neighborhoods either in proximate areas or further away. These efforts are now being supported by the federal government's Department of Housing and Urban Development, which has made resources available to support outreach efforts. In addition, many universities have turned to foundations and local businesses for financial support.

To illustrate the contribution of universities to building sustainable communities, this chapter offers an account of one university outreach project. Located in Baton Rouge, Louisiana, the project is known as Community University Partnership or "CUP." It is managed and jointly operated by a university task force of faculty, staff, and students at Louisiana State University (LSU) and a community organization known as the Metropolitan Community Housing and Development Organization. Its emphasis is twofold. First, it seeks to engage in community revitalization through a dynamic *partnership* between the university and local people. Second, it is committed to engaging in community development projects which build on local strengths to promote sustainable economic and social development and to engage in environmental restoration and protection projects. The chapter describes the CUP project, outlines its philosophy, and summarizes lessons that may be helpful to other universities and communities wishing to engage in sustainable community building.

CUP: THE EMPHASIS ON PARTNERSHIP

LSU's Community University Partnership is a collaborative effort between an interdisciplinary task force from the university and a community housing and development organization comprised of local residents, business owners, and religious leaders. LSU's main goal is to work in partnership with this organization to revitalize this declining urban community. The university's missions of research, teaching, and service are fulfilled through this project.

Target Area: South Baton Rouge

The area of engagement is a historically black area with a population of over 14,000 people. It is located between the downtown of Baton Rouge and the university. The boundaries of the community include the city center to the north, the Mississippi River to the west, the university to the south, and a prosperous neighborhood known as the Garden District to the east. The target area is known as South Baton Rouge because it formed the southern part of the city, before the university relocated into the area and the city expanded toward the south. The area is sometimes referred to as "The Bottom," a term originally used to describe the low-lying parts of the area that retain water after heavy rains. Formerly a hub of activity for the African-American community, this area now lacks many of the qualities that area elders remember fondly. In the first half of this century, locally owned businesses, schools, churches, and other social organizations were abundant and active in the area. These social institutions defined the community for residents and provided economic, social, and physical well-being for its residents (Dean, 1996).

Currently, the area is in a state of social, economic, and physical crisis, with its inner core presenting quite dismal statistics. Due to the proximity of the area to the university, the city's downtown, and the Garden District, the population characteristics of the core of the area are very different from those the block groups residing around the edges of the community. Many university students inhabit the southern ring, while the eastern and northern peripheries contain higher income individuals responsible for gentrification in certain subdivisions.

According to block group statistics from the 1990 census, 41% of those between the ages of 18 and 24 in the core area have less than a high school diploma. Fifty-two percent of the population lives at or below the poverty level, with over 29% living at or below 50% of the poverty level. Over 23% of households are headed by single parents. Furthermore, only 60% of households in the core have transportation. Ninety-two percent of the residents are African-American. The entire area reports more incidences of crimes per 1,000 residents than the city as a whole (Hurlbert et al., 1996).

The physical features of the area reflect its social and economic characteristics. Formerly busy streets are inundated with abandoned homes and other buildings. The total number of individuals residing in the area declined by a third from 1980 to 1990. Over 24% of housing units in the area are vacant (Hurlbert et al., 1996). Children in the area lack adequate adult supervision and are surrounded by desolate abandoned structures and vacant lots. The spatial isolation of the people is exacerbated by the fact that Baton Rouge's public transportation system does not reach all parts of the area and is extremely unreliable. In addition, vacant lots are frequently used to dump garbage and bulk waste materials, the streets are seldom if ever cleaned, and environmental pollution has become more serious.

The decline of the area has been associated with numerous factors. As in many other cities, the decline in employment opportunities for low-skilled workers associated with major changes in the American economy has increased the incidence of poverty (Wilson, 1987). Another factor has been the emigration of middle-class African-Americans from the area into the city's suburbs. Paradoxically, the civil rights movement, which ended legal racial segregation in the 1960s, allowed middle-class blacks to move into the city's more affluent areas. Another factor was the construction of Interstate Highway 10, which dissected the areas and required many businesses and residents to relocate (Dean, 1996). This cement interloper not only displaced hundreds of individuals but formed a wedge between residents in what was once a solidaristic community. It also exacerbated a growing problem of environmental pollution.

Forging the Partnership

While the area declined in population and lost vital local businesses, a core of committed residents remained, and many of those who vacated the area kept contact through kinship ties and church membership. In November 1993, a group of these community members came together to create an organization which it was hoped would address the area's pressing problems. The group was coordinated by the top aide of a state legislator who was raised in the community and is active in numerous civic and religious associations in Baton Rouge. Using his personal contacts, he mobilized a diverse group of concerned community members which included members of the clergy, long-time residents, homemakers, public housing residents, and local professionals including a physician, lawyer, and CPA, as well as several local business owners. These individuals met to discuss the area's problems. They decided to form an organization, known as the Metropolitan Community Housing and Development Organization (MCHDO), to address these problems. Although somewhat cumbersome, this name fits the organization's varied goals, which include economic development, the enhancement of educational opportunities, physical improvements, and the construction of housing. However, for most of the members, the slogan "take back" the community epitomizes their ideal of reconstructing a vibrant, sustainable community.

Concurrently with the formation of MCHDO, LSU began to expand its research capacity in the applied social sciences. The university embarked upon several activities to enhance research and service activities in these fields. This included a mandate to expand community outreach activities which would draw on the university's expertise in the social sciences. An associate vice chancellor for research and economic development was appointed and charged with developing opportunities for the university to engage in activities of this kind. The formation of MCHDO intrigued the newly appointed associate vice chancellor and he remained informed of developments through a social work intern who was placed in the office of the legislator whose aide initiated the project. The intern attended organizational meetings and assisted in MCHDO's development.

Recognizing the potential for collaboration, the associate vice chancellor began identifying faculty on the campus with an interest in community outreach. He formed an interdisciplinary task force of individuals already involved in community service activities and others who were interested in becoming involved. Thus, CUP was formed. In order to enhance the relationship that was developing between the university and the community, the social work intern was hired by the university. The purpose of her job was to facilitate the collaboration of the newly formed CUP Task Force and MCHDO, along with developing other human services research initiatives. She was to mediate the goals and objectives of MCHDO and CUP so that a partnership could be forged.

Those at LSU interested in enhancing the partnership recognized that there were major obstacles to overcome. Although LSU has abutted the area for over a century, it has not been a good neighbor to the local community. Historically, the university had no relationship with its African-American neighbors. During segregation, LSU was a white university and as such was off limits to blacks living within walking distance of the campus. Even after blacks were admitted to the university, they were not always made to feel welcome. Stories of LSU students driving through the neighborhood yelling racial epithets out of their car windows were not uncommon; nor were stories of blatant racism in the classroom. In addition, many community members were suspicious of the university and resentful of its lack of interest in their lives.

For this reason, much time and energy were spent on building personal relationships with MCHDO members and ensuring that a true partnership evolved. A great deal of effort has been devoted to formulating a viable working relationship which transcends a conventional service approach to community outreach. From the outset, it was agreed that LSU would not engage in projects that delivered services to "passive" community recipients but that all projects would be jointly identified, planned, and implemented by both the board and the CUP Task Force. LSU wished at all costs to prevent the emergence of a paternalistic service approach and wanted instead to enter into a dynamic dialogue about the problems facing the community and the best ways to respond to these problems. In this relationship, LSU offers its services to the community. It does not seek to impose projects or solutions on the MCHDO Board but instead believes that the board members know what is best for the local community. For this reason, CUP and MCHDO currently function as two autonomous bodies. However, while CUP and MCHDO are separate organizations with separate memberships, they collaborate closely. The chairman of MCHDO is invited to all CUP meetings and the director of CUP attends all MCHDO meetings. When appropriate, CUP Task Force members are invited to attend MCHDO meetings. This arrangement works well and has resulted in the development of a partnership which forms the basis for the project's operational approach.

The partnership can be viewed as both a goal in itself as well as a mechanism for fostering development. While university outreach is not new, true partnerships between community organizations and universities are rare. Forging a partnership is essential to gaining access to the community and ensuring that community-building efforts are sustained. Building a trusting relationship was the first important step in the process of breaking down barriers and gaining access to the community. The approach used was to rely on the connections of the legislator and the board president and the respect afforded to them by the community. Another aspect was to work slowly and consistently to gain people's trust. This approach was essential, as individuals from the university had worked in the community in the past, only to leave when their short-term projects had ended. The CUP Task Force aimed to create a permanent link between the university and the community that would outlast the individuals involved in the project, transforming the once-distant university into an accessible, sustainable community resource.

In addition to assisting in building relationships, the focus on the partnership provided an opportunity to assess individual and community concerns and desires without imposing any preconceived agendas. The overall framework of CUP developed and was strengthened through the dialogue that occurred during MCHDO meetings. In the early days of the project, the approach was simply to link MCHDO concerns and community strengths with LSU's expertise and abilities. As projects emerged and the dialogue broadened, the relevance of a developmental model to community revitalization became apparent.

USING A DEVELOPMENTAL MODEL

The developmental model adopted by the CUP project promotes the reciprocal nature of social and economic development. It rejects the idea that economic growth alone can promote the well-being of the population as a whole, and it criticizes traditional social programs for being impractical in times of slow economic growth. This approach is partly based on the writings of James Midgley (1995, 1996), who proposes a dynamic developmental model which implements programs that benefit individuals and communities in such a way as to help them to contribute to overall economic prosperity. Similarly, he proposes economic development strategies that produce sustainable social benefits for all.

The CUP project addresses these goals by implementing projects aimed at enhancing local economic activity through human capital development, social capital development, and programs that enhance productive employment and self-employment opportunities. All are intended to build sustainable communities through local effort. Individual projects will be discussed in terms of their relevance to these three types of community development projects.

Human Capital Development Initiatives

The first type of project implemented by CUP and MCHDO is designed to develop local human capital. Initiatives that build human capital enhance the abilities and opportunities of individuals through education, training, and improvements in public health and child welfare. This approach is based on the idea that economic development alone cannot promote the general welfare of a community. In areas plagued by distorted development, economic development alone does not provide ordinary people with access to opportunities because the rewards of this progress are shared by few. To avoid this bias in the distribution of the benefits of economic prosperity, investments in human capital are crucial. Human capital investment includes education, healthcare, nutrition, and child welfare. The CUP activities which exemplify investments in human capital are those which have established dynamic teaching and learning initiatives. These include literacy training, the oral history project, and service-learning initiatives.

Literacy Training

Literacy training is one example of a human capital investment. As individuals become literate, their opportunities for further education and employment dramatically increase. The elevation of literacy levels of an entire area increases the potential employment opportunities for the community by providing a work force able to undertake the duties required by today's jobs: thus, the area is able to attract new businesses.

The variety of literacy activities undertaken by CUP Task Force members include child literacy activities, family literacy, and adult literacy. Therefore, human capital is being developed in the full realm of individuals in the community. Members of the CUP Task Force from the Department of Communication Sciences and Disorders have recently completed a two-year literacy training program funded by the state of Louisiana. The purpose of the project was to determine the effectiveness of a "whole language" approach to family literacy training. As a result of the project, which ended in July 1996, adult participants gained up to two grade levels in reading recognition and comprehension for each 40 hours of instruction. The project is currently being extended through additional funding and community involvement.

In a separate initiative, the same faculty members received a renewal for a third year from the Louisiana State Department of Education to continue a preschool literacy program which they operate in the area. This funded an educational language development program that targeted two at-risk populations: children with identified speech and language difficulties and children from low-income neighborhoods. Students from the community functioned as peer models for those with language difficulties while receiving the benefits of intensive language and literacy learning activities. The preschoolers were recruited from the LSU Speech and Hearing Clinic and a housing project located in the South Baton Rouge community.

In addition, these faculty are engaged in developing instructional strategies for literacy training into a computer disk (CD-ROM) format, making literacy training more interactive. The instructional CD-ROM will enable literacy programs to train their staff with these innovative techniques.

Oral History for Human Capital Development

Oral history is a technique for recording previously undocumented history, often of marginalized groups. Recording the history of these people moves them out of the margins and helps them identify and access the qualities that promote their prosperity. When used as a method of education, oral history fosters human capital by teaching skills such as interviewing, documenting, public speaking, and leadership. These skills increase the participants' human capital by making them more competitive in the job market.

The CUP Task Force and MCHDO first ventured into the field of oral history by undertaking an oral history of McKinley High School. McKinley is the oldest African-American high school in the city and has a distinguished alumni. The oral history project began in the summer of 1995 and remained active throughout the 1995–96 academic year. The project was sponsored by the East Baton Rouge Parish Job Training Partnership Act (JTPA) summer work-based learning program and provided an opportunity for several students to engage in these activities. The students formed an Oral History Club at McKinley High School. The students continued their work during weekly meetings held either at McKinley High School or at the T. Harry Williams Center for Oral History at LSU. The students presented their findings to several education and English classes at LSU, as well as to local churches and community groups. The director of the T. Harry Williams Oral History Center, an assistant professor in the LSU College of Education, and graduate students from social work, history, and education assisted. The McKinley students also worked with a group of students in a service-learning class to write a script for a video of their presentation which was subsequently produced and is now available for sale. Proceeds are being donated to the McKinley High School Alumni Association to restore the old McKinley High School building, which is a registered National Historical Monument located in the heart of the South Baton Rouge community.

In addition to the video and community presentations, the students presented their findings at the Southwest Education Research Association Conference in New Orleans and to the American Education Research Association in New York in 1996. Both presentations were a great success and were received with great interest by those attending from other parts of the country.

Hoping to spread last year's success to more students and community members, the Oral History Project continued in the summer of 1996. It was again funded by the East Baton Rouge Parish JTPA and time donated by LSU. This time, the topic was the history of the South Baton Rouge business community. Four students recorded the history of business activity in this community. They compiled a document entitled *Pictures in My Head: South Baton Rouge Community Business and the Business Community,* which they presented at a well-attended community meeting. These students also gained valuable skills in systematically collecting information, preparing materials, and presenting their findings in public.

More recently, the oral history team received funding from the state of Louisiana to compile an oral history instructional package based on its experiences working with the local community. The project includes filming a video of oral history methods, writing an instructional manual of oral history techniques for teachers, and hosting a summer institute for teachers regarding the integration of oral history methods into teaching.

Service-Learning Initiatives

Service-learning is an innovative teaching modality which seeks to integrate students' learning in the classroom with a practical application of community service. Many disciplines have implemented service-learning classes into their curricula, and this learning technique is now widely used across the United States. At LSU, several CUP Task Force members have linked their service-learning classes with community projects identified with the MCHDO Board. Service-learning students have participated in human capital development projects such as math tutoring, assisting the oral history students in compiling a video, and creating a brochure containing information about literacy training and continuing education opportunities in the community. At the request of the principal of McKinley High School, CUP began an after-school program to tutor students who had failed certain sections of their exit exams. Both service-learning students and students under the direction of a professor in the engineering department are tutoring McKinley students in math and science. CUP Task Force members in the English department have also introduced service-learning into their composition courses. In the spring of 1996, an advanced English composition class linked its service-learning activities to CUP projects.

Social Capital Development Initiatives

Social capital development projects are the second type of development activities initiated by the CUP Task Force and the MCHDO Board to support their developmental approach to community building. These activities include (1) the creation and enhancement of social networks and social institutions that contribute to development, (2) the creation of community-owned amenities, and (3) the development of community and individual monetary assets (Midgley, 1995). All help to build sustainable communities.

Social networks and social institutions are social structures that support communication between individuals. They are studied by social scientists as entities separate from the individuals who form them. These forms of social capital have been shown to relate to economic activity in several different ways. For instance, social networks have been shown to facilitate employment opportunities for individuals (Granovetter, 1974). The density of social institutions has also been shown to contribute to the economic vitality of an area (Putnam, 1993). Midgley (1995) notes that community-owned institutions and structures also contribute to the economic stability of an area. Such facilities include schools, healthcare clinics, street improvements, and playgrounds, parks, and other community spaces. Finally, social capital can also refer to the monetary assets at the disposal of low-income communities. Individual Development Accounts (IDAs), as proposed by Sherraden (1991), are an example of this concept. IDAs give incentives to individuals to save money to help them provide for themselves and their families. Social capital projects initiated by CUP and MCHDO include a social capital assessment, the oral history project, the landscape architecture class project, environmental improvements, a community garden, and the renovation of Old McKinley High School.

Social Capital Assessment

In 1995, CUP Task Force sociologists were awarded a grant from the National Science Foundation (NSF) to undertake a study of the structure of social and economic isolation in underclass populations. This study involved a major survey of the community. The sociologists shared this idea with the MCHDO Board, which endorsed the project and facilitated the award of the grant by the NSF. The survey was undertaken during the fall of 1996. Community members were hired as telephone interviewers and field workers. Face-to-face interview participants were paid for completing the interview.

The assessment examined community social networks to study aspects of social and economic isolation in the community. The assessment first examined the presence of spatial mismatch, which is defined as a deficiency of low-skill employment opportunities in a low-income area. It then studied the types of social ties that exist among residents of the community with particular reference to the ways these ties are used to find employment. The survey not only provided a comprehensive perspective on the social capital currently existing in the community but will help identify ways of enhancing social capital development for economic change.

Oral History for Social Capital Development

While the oral history project was previously listed as a human capital project, it also cultivates social capital. The students who participate in the oral history project establish links with older members of the community, thus facilitating intergenerational communication. This not only strengthens the students' bond to the community, but enlarges the network of individuals that the student may access to find employment or educational opportunities in the future. Partially as a result of the ties developed during the Oral History of McKinley High School Project, two of the students will attend LSU in the fall of 1996. The other two students will attend college elsewhere.

Landscape Architecture Class

As noted earlier, the identification and development of infrastructure are key components in investing in social capital. In the fall of 1995, a graduate class of landscape architecture students worked with the community to assess needs and make recommendations about improving physical structures in the community. The class held several community meetings to receive input and share ideas about the needs and hopes of the community. At the end of the semester, the class held a final community meeting to present a variety of projects to community members and MCHDO Board members to use in planning and implementing changes in the neighborhood. The class presented a community map of the neighborhood which highlighted meaningful institutions and buildings within the community. They also presented the group with a comprehensive file of community resources to assist in future development initiatives. In addition, they drew sketches as examples of renovations that could revitalize dilapidated structures in the community. Finally, the students presented MCHDO with the idea of creating a community garden. This included information on how to begin a community garden project on vacant lots and how to start a public market with the products grown. While these various projects were primarily focused on the development of community amenities, they have positive implications for the future environmental enhancement of the areas as well.

Community Gardens

The landscape architecture class's recommendations about creating a community garden were accepted by the MCHDO Board and have resulted in the construction of the first community garden in the area. While the immediate goal of the garden may be to grow fresh produce to generate income, social capital can also be developed through the relationships and bonds which develop between neighbors who work in the gardens. In addition, the use of vacant lots for community revitalization contributes to the social capital of the area by converting community hazards into productive areas that make the area not only more visually appealing but economically viable.

Environmental Improvements

MCHDO believes strongly that the environmental deterioration of the area needs to be addressed, and it has undertaken two cleanups in the community. Each cleanup targeted a three- to four-block area and utilized block captains to facilitate the process. The first cleanup collected 14 tons of trash. The second collected over 50 tons. The municipal authorities were contacted to assist with the removal of the trash.

To further contribute to community beautification, MCHDO obtained a grant to plant trees along a frequently traveled intersection of a major road in the community. Large trees were planted along this road not only to improve the area but to absorb noise and fuel emissions from the highway.

Restoration of the Old McKinley High School Building

As noted earlier, McKinley High School is of considerable historic importance to the community. However, the school has moved to a newer building and its original structure has since been abandoned. The decay of the original building has dismayed many residents and graduates who are now collaborating to restore the structure. CUP has assisted with the project by using faculty from the university's School of Architecture and the Department of Construction to pre-pare estimates of the renovation costs and to assist with design plans. In 1996, the Louisiana legislature allocated funding to assist with the renovation of the building. It is hoped that the building will be used for incubating new businesses, housing non-profit organizations, and holding community meetings. The need for a small business incubator is strongly felt in view of the decline of small business activities in the area. Business incubators are facilities at which entre-preneurs are not only trained to start their own businesses but are supported and assisted on a routine basis until they are successfully established and able to function on their own.

PRODUCTIVE EMPLOYMENT AND SELF-EMPLOYMENT

To promote its developmental approach, the CUP project is also very much concerned with creating opportunities for productive employment and self-employment. Productive employment and self-employment interventions seek to facilitate income-generating activities which directly address the problems of material deprivation in low-income communities. They seek to increase partici-pation in the labor market through facilitating wage employment or supporting small business expansion or development. These activities are integral to pro-moting a developmental approach to building sustainable communities.

Job Link

This project, which is still in its formative stages, seeks to facilitate networks which will assist community residents in finding employment. While the Baton Rouge economy has recently been quite vibrant, many residents in the community are unemployed or underemployed. In 1996, a committee of individuals from both MCHDO and the university with knowledge of economic development projects and employment opportunities was appointed to address this issue. As a first step, the group collaborated with the local Job Services Office to implement a job information distribution network in the area which utilizes the local churches as distribution points of information. Plans are under way to develop a mentoring program whereby individuals seeking employment are assisted through the job search process by trained volunteers.

Small Business Support and Development

The employment strategy proposed by CUP is similar to the local economic development efforts being employed in many low-income neighborhoods. However, unlike the conventional approach which is widely used to train entrepreneurs and assist them to establish new small businesses on an individual basis, the CUP Task Force and MCHDO are seeking to strengthen existing businesses which are now under threat. In recent years, many locally owned small businesses have closed because of a lack of local support. The project seeks to promote community support for local enterprise by creating an awareness of the need to support local entrepreneurs. It also encourages the creation of local socioeconomic networks in which new enterprises can be nurtured and assisted by existing businesses. CUP and MCHDO are still seeking funding to implement this project.

ANALYSIS AND EVALUATION: LESSONS LEARNED

The CUP initiative has been in operation since the summer of 1994. After two and a half years, CUP and MCHDO have created a partnership to implement diverse projects that seek to implement a developmental model of sustainable community building. This project includes initiatives that build human and social capital and enhance opportunities for productive employment and self-employment. In December 1996, members of the CUP Task Force and MCHDO engaged in an examination of the project's achievements with regard to the partnership and its developmental projects.

The first goal of the CUP initiative was to develop a productive partnership between a predominantly white southern university and a predominantly African-American community. The partnership between the university task force and the community group is still developing. Although much progress has been made, more challenges must be faced. Although CUP has managed to convince a number of community residents that the university can be a responsible and productive neighbor, many other residents are not yet aware of or affected by these developments.

Nevertheless, the development of the partnership is CUP's most significant accomplishment given the history of the community and the university. Individual projects have contributed to this development by slowly building the trust of the community and allowing the CUP Task Force to prove its long-term commitment. LSU did not enter into this endeavor opportunistically because it had received funding to engage in community outreach. In fact, many of the projects were initially established through voluntary efforts, with funding being pieced together through existing resources or through limited external aid. When more significant funding was obtained by the university, MCHDO members were fully included in discussions of how the proposed project should be implemented. An example of this involved the planning and implementation of the NSF survey.

The second goal of the project was to implement a developmental approach to sustainable community building. The framework provided by this development model is holistic as it addresses social, environmental, and economic concerns. The integration of social, economic, and physical development is central to CUP's endeavors and includes human capital, social capital, and employment- and self-employment-generating projects. It also addresses environmental concerns and seeks to enhance the physical improvement of the area in conjunction with its economic and social development.

The CUP Task Force and MCHDO have been quite successful in implementing the developmental approach. As shown earlier, the various projects undertaken by the partnership have conformed to the general tenets of the model. This has been helpful because it has structured initiatives, given them a positive rationale, and ensured that sustainability remains a primary motive for the project. There is consensus among the CUP Task Force and MCHDO members that the model has strengthened the project and given it focus and direction.

In addition, the CUP Task Force and MCHDO members have learned several lessons which may benefit others engaged in similar activities. The first of these is that the history of the area in which outreach activities operate must be understood and issues arising out of this history must be addressed. The biggest obstacle that the partnership had to overcome was the negative perception of the university in the community and existing racial tensions related to the past. This history cannot be ignored and must be addressed before a partnership based on trust can be established. By dealing with this history, community members have accepted that CUP Task Force members are sincere and willing to deal with complex issues.

Another lesson is that a partnership is essential in building sustainable community projects. The partnership has been the key ingredient in making progress.

If they are to be successful, universities must transcend their often paternalistic role as service providers by developing a true partnership in which the full involvement of the community is ensured. This results in a truly reciprocal relationship. While the community can benefit from the involvement of universities, the universities can also benefit by using outreach activities to inform their research and teaching missions.

The CUP project has also demonstrated that successful community outreach projects take time to evolve and mature. It takes poor urban communities more than a few years to decline, and these communities cannot be rebuilt in a few years. Community building is a long-term process that involves forming coalitions among community members, associations, and groups. Building relationships takes time. It took two years just for CUP to become established. It will take many years of sustained effort before the long-term results of development projects will be fully appreciated. Unfortunately, many enthusiastic faculty and students are impatient and want to see quick results. CUP successfully resisted these tendencies by emphasizing the need for building strong relationships, investing in the future, and slowly building a sustainable program of action.

Finally, CUP has shown that while it is possible to begin a community initiative without money, funding can expedite the process. Working without outside resources actually enhanced the level of partnership achieved between the two groups since it required matching existing resources with community needs. Even so, sustainable community development requires funding. The gains made thus far would not have been possible without salary support for the CUP coordinator's position provided by LSU and the grant funding obtained by CUP Task Force members. This does not mean that nothing can be started without funding, but at least some funding commitments must be secured. In this regard, universities are well placed to assist local community groups to secure financial aid for projects which have a service as well as research and educational commitment.

The CUP project is still in its formative stages. Nevertheless, its current projects have already gone much further than was anticipated. Through developing a true partnership and adopting a viable theory of sustainable community development, it has already attracted attention and commendations from local people, civic leaders in the city, the university's central administration, and even faculty who are not directly involved. Hopefully, its experiences will be of interest to others in the academic world who believe that universities have a major role to play in promoting sustainable community building.

REFERENCES

Dean, P. (1996). Introduction. *Pictures in my head: South Baton Rouge community business and the business community.* Baton Rouge, LA: T. Harry Williams Center for Oral History, Louisiana State University.

Granovetter, M.S. (1974). *Getting a job: A study of contacts and careers.* Cambridge, MA: Harvard University Press.

Harkavy, I. and Pucket, J. (1994). Lessons from Hull House for the contemporary urban university. *Social Service Review, 68*(3), 299–321.

Hurlbert, J., Beggs, J., Shihadeh, E., Tolbert, C., and Irwin, M. (1996). *The structure of social and economic isolation in underclass populations.* Baton Rouge, LA: Department of Sociology, Louisiana State University.

Midgley, J. (1995). *Social development: The developmental perspective in social welfare.* Thousand Oaks, CA: Sage Publications.

Midgley, J. (1996). Involving social work in economic development. *International Social Work, 39*(1), 13–26.

Putnam, R.D. with Leonardi, R. and Nanetti, R.Y. (1993). *Making democracy work: Civic traditions in modern Italy.* Princeton, NJ: Princeton University Press.

Sherraden, M. (1991). *Assets and the poor: A new American welfare policy.* Armonk, NY: M.E. Sharpe.

Wilson, W.J. (1987). *The truly disadvantaged: The inner city, the underclass and public policy.* Chicago: University of Chicago Press.

Wilson, W.J. (1996). *When work disappears: The world of the urban poor.* New York: Alfred A. Knopf.

CHAPTER **8**

Making a Region Sustainable: Governments and Communities in Action in Greater Hamilton, Canada

Mark Bekkering and John Eyles

This chapter examines the nature and outcomes of a community stakeholder process to make sustainable development the philosophy to guide decision making in one of Canada's older metropolitan areas (see also Kendrick and Moore, 1995). The initiatives center on the political jurisdiction of the Regional Municipality of Hamilton–Wentworth (hereafter called "the region" or "Hamilton–Wentworth"), with a population of some 460,000 people of which over 70% live in the city of Hamilton. Hamilton–Wentworth is the upper-tier municipal government which includes six area municipalities. The region is the central planning authority for the purposes of physical, social, and economic planning and development.

Located in southwestern Ontario, along the western shores of Lake Ontario, Hamilton–Wentworth consists of a highly industrialized core surrounded by mainly rural communities. Approximately 40% of the region's land area is considered to be rural, with agriculture as the primary activity. The region encompasses a large number of environmentally significant areas, including the Niagara Escarpment, which has been designated as an internationally significant biosphere reserve by the United Nations. In 1994 Hamilton–Wentworth was selected by Environment Canada to receive, in the local government category, Canada's Award of Environmental Achievement.

The city of Hamilton, the dominant area municipality in the region, emerged as an industrial city of national significance in the mid-nineteenth century and remains the dominant force in Canadian steel making (Weaver, 1982; Dear et al., 1987). Its waterfront remains the focal point of steel, petroleum, petrochemical, and related industries, and over the years there has been severe pollution of

1-57444-129-9/98/$0.00+$.50

Hamilton Harbor (also known as Burlington Bay), which flows into Lake Ontario. While manufacturing industry remains significant in its contribution to the local economy, there has been a significant shift in employment structure to the tertiary sector, mainly government, education, and healthcare (Regional Munici-pality of Hamilton–Wentworth, 1992e).

We would argue that shift in employment structure has also brought about a shift in values (Eyles and Peace, 1990), from economic progress at any cost to a tempering of economic development with social and environmental con-cerns. In fact, one of the major growth industries in Hamilton–Wentworth focuses on environmental remediation and pollution prevention. This then is the backdrop against which the attempts to make a sustainable community may be viewed.

INITIATING THE SUSTAINABLE COMMUNITY PROJECT

The poor quality of Hamilton's natural environment was documented through-out the 1980s with reports on poor air quality and its effect on health (Regional Municipality of Hamilton–Wentworth, 1990d, 1994a; Eyles et al., 1996) and the quality of the water in Hamilton Harbor (Canada, 1991). The International Joint Commission, a binational body established by the United States and Canada to determine and suggest remediation strategies for Great Lakes water quality, identified over 43 areas of concern in the Great Lakes Basin, one of which was Hamilton Harbor (Health Canada, 1992). Part of the proposal for remediation was the development of a Remedial Action Plan (RAP) initiative, a multistakeholder round table approach for plan development, involving rep-resentatives from the community, industry, government, and non-government organizations (including environmental groups). The RAP showed that organi-zations with divergent views and interests could work together to address community problems. It became the model for Hamilton–Wentworth's Sustain-able Community Initiative. It framed the political response as much as the environmental degradation fueled the need for action. The importance of civic involvement to address problems has, of course, been recognized by others (see Putnam, 1993; McKnight, 1995).

Why Sustainability?

The Sustainable Community Initiative was a creation of the regional govern-ment, and it began in 1989, when the region's senior management team decided that new mechanisms were needed to improve the coordination between munici-pal budget decisions and policy goals and objectives. At the same time, the region's official plan and economic strategy were in need of comprehensive review, and questions were raised about what directions or philosophies would guide this update process. Linked to this was the recent election of a regional chair who had campaigned on a platform of addressing issues of environmental protection, the need for more affordable housing, and opening up the decision-making process of government. His platform for election was in part a reflection of the concerns being expressed by members of the community through their heightened awareness and concern about the environment.

The Planning and Development Department, responsible for the development of the region's official plan and economic strategy, was mandated with identifying a guiding philosophy for addressing these concerns. After research and deliberations, the region's management determined that sustainable development would be an appropriate guiding philosophy (Regional Municipality of Hamilton–Wentworth, 1989).

A Citizen Task Force

Based on this recommendation of management and the research of the Planning and Development Department, Regional Council in June 1990 created a citizen task force. This group of 18 volunteers, officially called *The Chairman's Task Force on Sustainable Development,* was given a two-year mandate to complete the following six tasks:

- To develop a precise definition of what sustainable development means to Hamilton–Wentworth, to be used in developing an overall vision for the region
- To develop a community vision to guide future development in Hamilton–Wentworth based on the principles of sustainable development
- To establish a public outreach program to increase awareness of the concept of sustainable development and to act as a vehicle for feedback on potential goals, objectives, and policies for the region
- To provide input as to how the concept of sustainable development could be turned into practical applications through regional initiatives
- To demonstrate and articulate in detail the usefulness of the sustainable development concept in the review of the region's long-term planning policies
- To provide direction to staff and the Economic Development and Planning Committee, which would be using the concept to guide its review of the region's economic strategy and official plan (Regional Municipality of Hamilton–Wentworth, 1990a)

To assist the task force in fulfilling its mandate, a project team consisting of one full-time coordinator, one full-time communications advisor, and six part-time researchers was put in place. A technical advisory committee was also created to assist the task force in designing its public outreach program. The technical advisory committee was made up of regional staff and representatives from organizations who had experience in community development efforts.

To facilitate understanding within the community and Regional Council about the purpose of the initiative, it was linked to the development of the region's official plan and economic strategy. The efforts of the task force and its final reports were to shape these two documents which guide the decision making of Regional Council. To formalize that link, the chairman of the task force was the chairman of Regional Council's Economic Development and Planning Committee, and all reports were first presented to this committee before going to Regional Council.

A Multistakeholder Approach

To engage the community, the task force was established as a multistakeholder round table (see National Round Table, 1993; Canadian Standards Association, 1996); it originally consisted of 18 people representing government, academia, industry and commerce, environmental organizations, resident associations, health organizations, women's groups, the development industry, arts and cultural groups, the business sector, agriculture, and the real estate sector. Potential members for the task force were asked to submit an application identifying their interest in the effort and what they would bring to the task force. An interview committee consisting of the chair of the task force, who was a member of Regional Council, and the project coordinator selected the members of the task force. The chairman of the task force interviewed 50 applicants and selected members based on their (a) knowledge and experience in at least one of the sectors of interest, (b) willingness and ability to work with different groups and organizations, (c) ability to see a broad rather than a narrow perspective, and (d) commitment to work with other members of the task force.

The selection process did target and identify key constituencies. For example, the two major steel manufacturers in the community have a large influence on the local economy and natural environment. Therefore, it was seen as crucial that they had an official representative on the task force. In each of the sectors identified as requiring representation, if no suitable candidate volunteered, a search effort was made to identify and encourage desired representatives to participate in the task force. Some groups were not represented because they declined to participate (such as the local school boards) or were missed by the organizers (such as youth) or were dealt with through the public consultation process (such as people residing in nursing homes and people living in emergency shelters).

All of the citizens selected to sit on the task force were very committed to and involved in the community. Most already had some understanding about sustainable development and all definitely knew what the major issues facing the community were. To ensure everyone had some common understanding, initial meetings focused on discussing and developing a definition of sustainable development (see Block, 1987). In some respects, the concern with definitions focused attention on process (agreement over terms, planning directions) rather than actions. And indeed the goal of this initiative has been to change the planning parameters of the regional government.

Being created under the authority of the regional chairman, all reports and recommendations were first presented to the regional chairman. It was then his decision as to whether they would be taken and presented to Regional Council. While he agreed with the recommendations of the task force, the process of planning a sustainable community can be seen as being guided by the existing political authorities and structures.

Although the members of the task force designed their own work and public outreach program, they did not have any direct control over financing. Essentially they decided upon what they would like to do, and the project staff worked within the budget of the Planning and Development Department to make their wishes occur. Having a relatively flexible budget is one of the major reasons that contributed to success. No serious financial restrictions were placed on the task force as it went about its mandate. The only significant restriction placed on the effort was time. Regional Council was reluctant to extend the time of the mandate and pressed the task force to complete its efforts by the end of two years. Regional Council did, however, eventually give the task force a six-month extension on its mandate.

The multistakeholder round table approach was very successful in making a greater number of people more comfortable in participating in this form of community-led decision making. The approach has since been used by the region in a number of other initiatives, such as watershed planning exercises and a community review of municipal government structure in Hamilton–Wentworth.

Consulting the Community

The terms of reference for the Chairman's Task Force on Sustainable Development required, as one of its six purposes, the establishment of a "...*public outreach programme to increase public awareness of the concept of sustainable development and to act as a vehicle for feedback on potential goals, objectives and policies for the Region*" (Regional Municipality of Hamilton–Wentworth, 1990a). There were nine goals to the public outreach program which fell into the three general categories of education, citizen input, and quality. The goals were:

- To inform the general population of the basic principles of sustainable development and of the purpose and mandate of the task force
- To inform citizens of the range of regional government activities, such as public expenditures and investment, the regional official plan, and the economic strategy
- To communicate information generated by citizens back to the public
- To gather citizen concerns and perceptions regarding the quality of their environment and life that can be used to identify issues

- To gather citizen perspectives on basic values and goals that can be used to develop a set of principles to guide the preparation of a regional vision statement
- To reach out to groups in the population that are not normally part of the decision-making process, such as children, youth, the disadvantaged, and the non-English-speaking community
- To develop community awareness and support for the work of the task force that will result in long-term community support for the implementation of the regional vision statement
- To achieve meaningful citizen participation that provides good quality information to the task force and is an empowering exercise for citizens
- To draw out those citizens who wish to be involved more deeply in the task force's work as members of issue working groups

To achieve these goals, five strategies were employed: a media campaign, individual feedback opportunities, community workshops, focus groups, and a large community forum. Over the two and half years of the task force, almost a thousand citizens were eventually involved directly or indirectly in the work of the task force.

IDENTIFYING COMMUNITY ISSUES

In the fall of 1990, the task force initiated the first three parts of the public participation strategy. A broad media campaign was conducted that included the use of local print, radio, and television media; the development and delivery of information booths in local shopping malls and other locations; and the preparation and distribution of 150,000 copies of the task force newsletter. The campaign informed the community about the purpose of the task force and the upcoming opportunities available to people to become involved in the work of the task force.

The major activity held at this time was the seven community workshops or town hall meetings in the different municipalities in the region. Approximately 160 people participated in these sessions designed to identify which issues needed to be addressed and which values should guide the work of the task force. At these town hall meetings, which were facilitated by members of the task force, people were asked to discuss: (1) what they liked about life in Hamilton–Wentworth, (2) what detracts from life in Hamilton–Wentworth, (3) what should be done to improve life in Hamilton–Wentworth, and (4) what values they felt should guide decision making in Hamilton–Wentworth.

In essence, the purpose of the questions was to elicit citizen thoughts on what issues need to be addressed and which values should guide the ongoing work of the task force. The results of each town hall meeting were summarized in a report provided to each member of the task force and made available upon request to the public (Regional Municipality of Hamilton–Wentworth, 1990b). People who did not wish to participate in the town hall meetings were provided the opportunity to contribute their ideas by submitting written comments or phoning the Ideas Telephone Line. The task force sponsored a series of focus groups for people normally overlooked in public outreach programs. The questions discussed were the same as those used in the town hall meetings. Approximately 150 people participated in 18 focus groups. Organizations participating were Ancaster Senior Achievement Centre; Citizens Action Group; Elizabeth Fry Society; Good Shepherd Centre; Housing Help Centre; John Howard Society; Department of Public Health Services; Kirkendall–Strathcona Neighbourhood House; Promoting Elders' Participation Project, Victorian Order of Nurses; Social Planning and Research Council; Kathleen Ward's Second Level Lodging Home; Native Women's Centre; Neighbour-to-Neighbour; People First; Polish Immigrant Services; Project First Step; and Sexual Assault Centre (Regional Municipality of Hamilton–Wentworth, 1990c).

The project staff under the guidance of the task force assembled all of the information provided in these different forums and analyzed it using simple content analysis. All of the ideas and concerns were organized into groups and rank ordered according to the frequency of times they were commented on in the different forums.

Reviewing the community input, the task force identified seven major issues of concern in Hamilton–Wentworth. In no order of priority, these were:

- Transportation system offers inadequate opportunities for cycling, pedestrians, and public transit
- Recent urban development is unattractive, destructive of landscape character, wasteful of resources, and lacks a sense of community
- Pollution of air, water, and soil
- Loss of natural areas and encroachment on conservation lands and scenic areas
- Economic concerns regarding overdependence on manufacturing, a lack of dynamic initiative in the economy, and shrinking employment opportunities
- Local government is not responsive to citizens, shows a lack of leadership, and exhibits little commitment to long-term plans and policies
- Social problems such as poverty, security, and an aging population

In addition, nine specific values were commonly expressed by the community. These values are best expressed as directives of the kind of regional community people would like to see. In no order of priority, they were:

- Ensure community character and identity are preserved and enhanced
- Preserve and enhance natural areas and amenities
- Preserve farmland and the rural landscape

- Ensure the continuance of a friendly, safe, and diverse human community
- Develop an integrated, balanced, and efficient regional transportation system
- Protect and rehabilitate the air, water, and soil
- Develop a self-sufficient, diverse, sustainable local economy
- Improve the appearance and fit of the built environment with the natural and community context
- Alleviate poverty

DEVELOPING A COMMUNITY VISION

The next phase of the work of the task force focused on the development of a community vision which would be based on the concept of sustainable development and reflect the issues of concern identified in the first phase. The key elements of this stage of the process were (1) preparation and presentation of 11 discussion papers, (2) organization of eight citizen vision working groups, (3) a large community forum, (4) preparation of the first draft of the community vision, (5) community review of that vision, and (6) presentation of the final vision statement to Regional Council.

To assist the task force in reviewing the issues and assessing their importance, the Planning and Development Department, with the assistance of a number of other regional departments, community organizations, and researchers from McMaster University (the local university), prepared 11 discussion papers (Regional Municipality of Hamilton–Wentworth, 1991a). These papers were:

- Government in Hamilton–Wentworth
- Demographic Trends and Social Adjustment
- Workforce Education and Human Development
- Transportation, Physical Services and Land Use
- Economic Base and Livelihood
- Food, Rural Land Use and the Agricultural Economy
- Environment and Health
- Greenspace and Natural Areas
- Energy, Waste and Resource Consumption
- Poverty, Social Equity and Community Well-Being
- Population Health and the Health Care System

The papers reviewed past trends and highlighted issues which may be of concern in the future. They made no recommendations on possible solutions but rather focused on reviewing the trends and trying to relate them to the concept of sustainability. Research and evidence for decisions concerning the vision were then key components in developing the strategy for a sustainable community.

VISION WORKING GROUPS

Working with the issues and values identified by the community and new ones identified from the discussion papers, the task force organized eight volunteer Citizen Vision Working Groups. The mandate given to the working groups was to examine in detail an assigned topic area and develop a vision statement for that topic area. The topic areas were human health; ecosystem integrity; natural areas; community design; culture and learning; community well-being; economy, livelihood, and education; and food and agriculture. The working groups consisted of 35 citizens who volunteered and tended to be mainly people who were unsuccessful candidates for the task force. The working groups which occurred in the winter/spring of 1991 met on average once every week or two weeks and were chaired by a member of the task force. The volunteers conducted all of their own research and in most cases prepared their own final reports (Regional Municipality of Hamilton–Wentworth, 1991b).

Community Forum

When the working groups finished their work, a one-day community forum called "Creating the Sustainable Region" was held in June 1991. The purpose of the forum was to review the draft reports of the eight vision working groups and to allow people the opportunity to contribute their views on the future. Attended by 250 people, the community forum was free and open to whomever in the community wished to attend.

After the working groups reported the results of their workshops, 18 business and community organizations made presentations to the task force about their vision for the region's future (Regional Municipality of Hamilton–Wentworth, 1991c).

A Community Vision

The task force then worked to produce a consensus summary of what type of community people would like to live in. This led to the first draft of the vision statement, *VISION 2020: The Sustainable Region.* The vision states that sustainable development is positive change which does not undermine the environment or social systems on which people depend. It requires a coordinated approach to planning and policy making that involves public participation. Its success depends upon widespread understanding of the critical relationship between people and their environment and the will to make the necessary changes. Principles of sustainable development encompass the following:

- Fulfillment of human needs for peace, clean air and water, food, shelter, education, and useful and satisfying employment

- Maintenance of ecological integrity through careful stewardship, rehabilitation, reduction in wastes, and protection of diverse and important
  natural species and systems
- Provision for self-determination through public involvement in the
  definition and development of local solutions to environmental and
  development problems
- Achievement of equity with the fairest possible sharing of limited
  resources among contemporaries and between our generation and that
  of our descendants

These basic values underlie *VISION 2020*. The vision expresses ideas contributed by citizens through several phases of community participation. It is the
beginning of an ongoing process leading to a sustainable region (Regional
Municipality of Hamilton–Wentworth, 1992b).

The first draft of the vision statement was included in 150,000 copies of the
task force newsletter, which was distributed to every household in the region in
January 1992. The newsletter informed people of how they could comment on
the vision and recommend any changes to the vision they would like to see.
People were invited to either submit written comments or attend a community
meeting held in March 1992.

Over 50 written recommendations from a wide variety of individuals and
organizations were submitted to the task force and included in its regular monthly
meeting agenda packages. Attended by 65 people, the community meeting
included eight verbal presentations, an additional three written submissions, and
an open discussion about the content of the first draft of *VISION 2020* (Regional
Municipality of Hamilton–Wentworth, 1992a).

The initial release of the vision statement resulted in negative coverage in the
local media. The vision statement was seen as being too short and vague to have
any value for decision making. The task force was criticized for producing a
document which contained little specific directions or actions for addressing the
issues of concern in the community. Taking their cue from the media, the
majority of the public also responded very negatively to the vision statement.
Yet the task force had decided that it would only present a vision statement at
this phase of the project because it wanted to ensure that it represented the goal
that people wanted to achieve. Recommendations for how to achieve that vision
would come from the next phase of its mandate.

The task force was not swayed by the criticism in the media and decided to
continue with the approach it had mapped out at the start of its mandate. Task
force members examined the written submissions and the verbal presentations
made at the community meeting and revised the vision statement. It remained
a four-page document that describes Hamilton–Wentworth in the year 2020. The
year 2020 was selected because it put forth almost a 30-year time frame for
achieving the vision, which coincided with the usual time frame for the development of the region's official plan.

The vision statement was adopted by Regional Council in June 1992 as the basis for regional decision making in Hamilton–Wentworth, including the official plan, regional economic strategy, and the capital budget process. The vision statement had not only an awareness function but a direct impact on economic development and investment decisions.

DEVELOPING A PLAN OF ACTION

To help guide actions to be in keeping with the sustainable community concept expressed in *VISION 2020,* the task force developed a set of recommendations which identified the types of decisions, actions, and policies required by government, community groups, businesses, and individuals if *VISION 2020* is to become a reality for the community.

Work on identifying recommendations for action started in spring 1992 and involved the creation of eight teams of volunteers organized around specific topic areas: agriculture, rural settlement, and the rural economy; economy, livelihood, and work force education; community well-being, health, and quality of life; waste management, physical services, and urban growth; transportation; land use planning and community design; cultural, historical, and recreational resources; and natural areas and natural resources. Called *implementation teams,* these groups were charged with the responsibility of reporting to the full task force on the decisions and actions for reaching the assigned areas of *VISION 2020.* Almost 75 people participated in the implementation teams and the development of their final reports.

The people who participated were a mix of the citizen volunteers from the original vision working groups and invited individuals from various organizations and government departments in the community. The goal was to mix the people who had developed the vision with people whose agency may take the lead role in making the vision a reality. For example, the Transportation Implementation Team consisted of the former working group members, people from the region's Public Transit and Transportation departments, people from Accessible Transportation Services, and individuals from cycling organizations.

Similar to the vision working groups, the implementation teams conducted their own research and prepared their own reports. Each implementation team was chaired by a member of the task force and provided with a staff person to assist in identifying research materials, organizing meetings, and taking minutes. These groups completed their reports in August 1992 (Regional Municipality of Hamilton–Wentworth, 1992c). The amount and level of detail presented in the implementation team reports vary according to strength and commitment of the volunteers who participated in the teams. Once the implementation teams had completed their reports, an all-day community workshop entitled "Creating the Sustainable Region" was held in September 1992. The purpose of the workshop was to review the implementation team reports and to allow citizens the opportunity to contribute their ideas for implementation. Approximately 200 people attended and reviewed the reports in small workshop settings (Regional Municipality of Hamilton–Wentworth, 1992d).

The task force recognized that it did not have the time left in its mandate to develop a detailed action plan which defined responsibilities, time frames, and required financial resources. A decision was made to present two reports to Regional Council. The first, entitled "Directions for Creating a Sustainable Region" (Regional Municipality of Hamilton–Wentworth, 1993a), provides greater detail to the vision statement by outlining more specific goals that the community should set to achieve the vision. The second, entitled "Detailed Strategies and Actions" (Regional Municipality of Hamilton–Wentworth, 1993b), is essentially a summary of the types of decisions and actions required to implement the goals and vision statement.

ACTIONS TO CREATE A SUSTAINABLE COMMUNITY

The issues raised by the community and reflected in the final reports of the Chairman's Task Force on Sustainable Development can be summarized into 12 major areas of concern. The policy shift areas and examples of the 400 detailed recommendations are listed below:

- Identifying and protecting natural areas and corridors (example: in cooperation with all stakeholders, identify the hierarchy of natural areas and corridors, together with policies controlling the land uses within and around the system of natural areas)
- Improving the quality of water resources (example: implement a user pay concept by metering all water users)
- Improving air quality (example: develop a minimum standard for the amount of vegetation required on residential lots)
- Reducing the amount of waste being produced and going to landfill (example: establish a waste exchange depot)
- Reducing energy consumption (example: continue the conversion of public transit vehicles to natural gas)
- Creating a more compact and diverse urban form (example: in the region's official plan designate a firm urban boundary beyond which urban development will not be permitted)
- Changing our mode of transportation (example: purchase public transit vehicles that can accommodate wheelchairs)
- Ensuring good health for all and adequate social services (example: undertake efforts to ensure accessible, affordable, nutritious, and personally acceptable supply of food, safe drinking water, and housing for everyone)
- Supporting the arts community and recognizing cultural diversity

- Empowering the community (example: hold regular town hall meetings and other forums to facilitate citizen input)
- Diversifying the local economy, provision of appropriate training, and supporting environmental companies (example: create a centralized resource center to assist people wishing to start a business)
- Supporting the local agricultural sector (example: permit the direct sale of farm produce to the public)

The task force recognized that it did not have the time or probably the resources to fully investigate all possible actions required to implement the vision. Its two final reports are therefore guidelines for action. The strength of these two final reports is that they are consensus documents. All 400 goal statements and recommendations presented were agreed to by every member of the task force.

In January 1993, an estimated 300 people gathered at the Hamilton Convention Centre in downtown Hamilton for the presentation by the task force of its final reports to the region's Economic Development and Planning Committee. The reports were adopted by the Economic Development and Planning Committee and eventually, with a unanimous vote in February 1993, by Regional Council.

When the final reports of the task force were presented, they received wide public support. The response in the media was extremely positive, and numerous reports were presented congratulating the members of the task force and discussing the challenge of implementation. The initiative did not, however, result in a detailed action or implementation plan. In part, the overriding goal of the task force was to build community commitment to a set of goals around a sustainable community.

IMPLEMENTING THE PLAN AND THE VISION

Before the task force completed its mandate, a lot of time was spent discussing the mechanisms required to facilitate implementation of *VISION 2020*. Two methods considered were the creation of a citizen organization with the mandate to encourage and facilitate implementation throughout the community and the development of a similar group of regional staff who would recommend to Regional Council how it could implement the vision statement.

Former members of the task force with the assistance of the project staff tried to create a citizen's organization called Citizens for a Sustainable Community. Although it still exists as an organization, it has never been able to attract the attention of the community and build upon its original membership of around 50 individuals. To a large extent, the organization has had essentially little impact on the community and has been able to facilitate very little implementation of any aspects of the vision statement. Its one major achievement has been involvement in the Young Citizens for a Sustainable Future program. Developed by representatives of the Environmental Health Program at McMaster University and the region, this is a leadership training program for secondary school students that helps them understand the concept of sustainability and the trade-offs required in local government decision making (Regional Municipality of Hamilton–Wentworth, 1995). The Citizens for a Sustainable Community played a lead role in finding original financial resources and many of the volunteer advisors for this program.

Within regional government there has been a more coordinated effort to implement the vision statement, as diagrammed in Figure 8.1. After approving the vision statement and implementation recommendations, Regional Council organized the Staff Working Group on Sustainable Development. This committee of senior staff from every department has been mandated with the responsibility of facilitating the implementation of the vision statement in the operations of the municipality. This group was organized in 1993 and is still in operation.

The working group has identified, recommended, assisted, and guided the implementation of a number of initiatives that have attempted to integrate the concept of sustainability and the goals of the vision into the municipality's decision-making process. The most significant has been the *Sustainable Community Decision-Making Guide* (Regional Municipality of Hamilton–Wentworth, 1996a). The guide is a simple process to be used by all regional staff in evaluating all proposed projects to ensure a balance has been reached among environmental, economic, and social/health factors and that the goals of *VISION 2020* have been considered. In August 1994, Regional Council directed staff to use the *Sustainable Community Decision-Making Guide* as a tool to assist in the evaluation of all proposed and existing policies, programs, and projects. Now all reports presented to Regional Council must identify the possible sustainable community implications of the recommended action.

In June 1996, Regional Council mandated the Staff Working Group to begin work on revising decision-making procedures to incorporate the concept of sustainability in the areas of grant applications, interview and candidate selection for citizen advisory committees, tendering and purchasing policies, and internal auditing procedures.

POLICY DEVELOPMENT

The second major effort of Regional Council has also focused on integrating the goals of *VISION 2020* into the decision making by revising long-range planning and policy documents. Reflecting one of the original goals of the project, the region's official plan was completely revised and renamed *Towards a Sustainable Region* (Regional Municipality of Hamilton–Wentworth, 1994b). Adopted by Regional Council in June 1994, the official plan incorporates directly almost

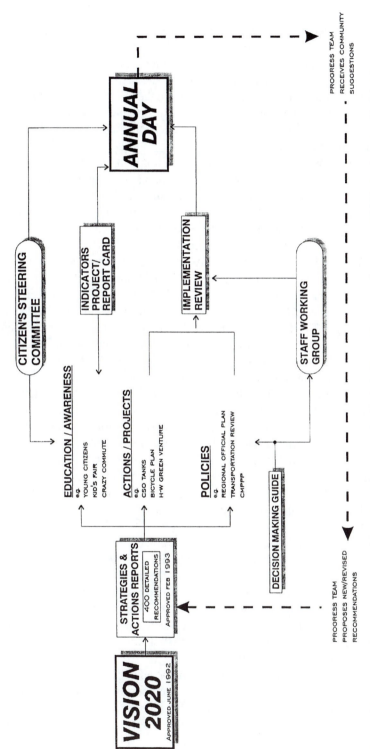

Figure 8.1 Hamilton–Wentworth *Vision 2020* sustainable community initiative. (From *Sustainable Community Project*, Regional Environment Department, T&C Section, Regional Municipality of Hamilton–Wentworth, 1997.)

100 of the 400 detailed recommendations made in the *VISION 2020* reports. Because the official plan deals primarily with land use issues, the recommendations made by the task force around encouraging a more compact and mixed urban form and developing systems to encourage alternative transportation choices and protect environmentally sensitive areas were the key directions incorporated into the plan. The official plan is approved by the province of Ontario's minister of municipal affairs and is used to guide decision making related to future development in the region.

If Regional Council chooses to make a decision that contradicts the policies stated in the official plan, the citizens of the community have the opportunity to appeal that decision to the Ontario Municipal Board (OMB), the supreme planning court in the province. There have been some incidents where members of the community have had a decision of Regional Council changed to one which was deemed to be more in line with the goals of *VISION 2020*. For example, in 1995 the Town of Dundas, a municipality in the region, proposed allowing one-acre lot development in an area deemed to be of some environmental significance. The Conserver Society of Hamilton and District opposed this decision on the basis that it was a decision that did not fit with the goals of *VISION 2020* and the region's official plan. The OMB ruled in favor of the Conserver Society and stated in its decision that it agreed that the proposed development plan was not consistent with the community vision (Canada Law Book, 1996).

Another significant long-range planning effort was the development of the regional transportation review. Completed in late 1996, the transportation review outlines the transportation services and policies required to achieve the transportation goals of *VISION 2020*. It provides direction on how to balance regional spending between roads, parking, public transportation, and pedestrian and bicycling facilities. This report has met some resistance from the community, particularly those who are hesitant to make any changes in the existing road system that may reduce convenience for automobile users. Therefore, many of the recommendations of this report have yet to be approved by Regional Council.

Initiated in 1994, another major long-range planning exercise was the Comprehensive Municipal Pollution Prevention Plan (Regional Municipality of Hamilton–Wentworth, 1996c). The focus of this project is to involve regional staff in developing an action plan, for the region to incorporate pollution prevention into its daily operations. The goals were to: (1) make sure "our own house" is in order, (2) look at how the region can influence pollution prevention initiatives through its regulatory authority, and (3) share project results with other communities.

The Pollution Prevention Project formally ended in 1996, but continues through various employee committees that are constantly evaluating and suggesting changes in regional operations so that there is a reduction in impact on the natural environment. Building upon this effort, the region has recently initiated work on developing an environmental management system with the desired goal being registration under the International Standards Organization 14001 series.

The final major long-range planning exercise was the development of an economic strategy entitled *The Renaissance Project*. Prepared by a group of business leaders and with the assistance of over 100 citizens from many different sectors of the community, this report was adopted by Regional Council in November 1994 as its strategic plan for long-term economic development. The goals of the strategy are to: (1) create societal wealth, (2) create jobs or other meaningful means of livelihood, and (3) incorporate sustainable development principles into community economic development activities.

In addition to the development of long-range planning documents and policies, Regional Council has initiated a large number of specific initiatives to address the detailed concerns raised in the *VISION 2020* reports (see Table 8.1).

COMMUNITY IMPLEMENTATION

In the broader community, implementation of *VISION 2020* has been more sporadic. No company or community organization has officially endorsed *VISION 2020*. However, events and activities sponsored by the region have received excellent support from the community and corporate sponsorship. Over 200 local organizations and businesses have been involved in events such as the Children's Sustainability Fair, the Annual *VISION 2020* Sustainable Community Day, and the Young Citizens for a Sustainable Future program.

Local industries and community organizations have made a limited effort to relate their activities to the goals of *VISION 2020*. The majority of these organizations are, however, making significant efforts in the area of environmental protection, but not because of any commitment they have made to *VISION 2020*. For example, 15 of the major industries in Hamilton have recently formed a coalition to develop a plan for reducing their cumulative impact on the natural environment. This task group has been arranged to create a partnership to address concerns being expressed by the community regarding pollution from these industries. While the region is an invited partner, the initiative started with one of the major steel producers. The underlying goal of this initiative is not to implement *VISION 2020* but rather to address community concerns, improve profits, and improve community relations. Although an effort that is in keeping with *VISION 2020,* it is an initiative that the organizers did for other reasons.

VISION 2020 remains largely as a guide for the decision making of Regional Council as opposed to the decision making of everyone in the community. The community-led review process that has been proposed for 1998 and is currently being developed by regional staff will try to address the issue of creating stronger community ownership.

Table 8.1   Actions Initiated by Regional Council to Implement *VISION 2020*

Natural areas
- Regional Greenlands Strategy[a]
- Environmentally Significant Areas Impact Evaluation Group[a]
- Red Hill Creek Watershed Plan
- Roadside and park naturalization[a]

Water and wetlands
- Distribution of water conservation kits
- Construction of combined sewer overflow tanks[a]
- Involvement in waterfront park development

Transportation and air quality
- Hamilton–Wentworth Air Quality Initiative[a]
- Regional Bicycle Commuter Project[a]
- Taxi Scrip Program
- Conversion of buses to natural gas
- Purchase of wheelchair-accessible buses

Energy and waste management
- Support for the Hamilton–Wentworth Green Venture[a]
- Pollution prevention plan[a]
- Energy management project[a]
- Development of an environmental management system
- 3 R's Programs

New economy
- Greater Hamilton Technology Enterprise Centre[a]
- Environmental commitment awards
- Employment services and support

Well-being and health
- Kidestrian
- School Nourishment Task Force[a]
- Neighborhood community capacity-building initiatives
- Non-smoking bylaw

Monitoring
- State of the Environment Reporting
- Fact book on health
- Sustainable community indicators

Education, awareness, and empowerment
- Annual *VISION 2020* Sustainable Community Day
- Children's Sustainability Fair
- Young Citizens for a Sustainable Future
- The Constituent Assembly
- Open houses and presentations

[a] Major projects started because of *VISION 2020*.

HEIGHTENING AWARENESS OF THE SUSTAINABLE COMMUNITY

While having limited response from the community, the Sustainable Community Initiative has heightened awareness about sustainability. As part of ensuring community recognition and ownership of the process, an annual Sustainable Community Day was initiated in 1993 and has been run every year since that time. The event serves as a mechanism to bring the community together to discuss progress and future priorities. Part of this event, since 1996, has been an annual report card based on a set of sustainability indicators, developed by the region and other research partners but through a process that encouraged community participation so that the indicators would be understandable, realistic, motivational, and credible in the eyes of the community. Working through different forums of public input, 29 indicators were finally selected. They were presented to Regional Council in summer 1996 and approved for use in monitoring broad community progress in relation to the goals of *VISION 2020*. The 29 indicators were presented to the community at the Third Annual *VISION 2020* Sustainable Community Day in October 1996 in a report card format. The trends for 1993 to 1995 for each indicator were evaluated and rated according to one of the three following categories: needs improvement, hard to say, and making progress (Regional Municipality of Hamilton–Wentworth, 1996b).

Ongoing monitoring of the indicators is the responsibility of the Strategic Planning Division of the Regional Environment Department, and a report card will be prepared for presentation at every Annual Day. Whether progress is being made will be evaluated by the community at the Annual Day. Linked to the indicators project is the *VISION 2020* Implementation Review, which is currently being developed and should be completed in time for its first presentation at the Fourth Annual Day in 1997. When complete, the review will provide a detailed examination of the 400 recommendations made by the Task Force on Sustainable Development. The review is identifying actions made to implement those recommendations, actions that may be considered as contradicting the recommendations, why some recommendations have never been acted upon, and where there may be new issues that were not identified in the original development of *VISION 2020* and the detailed recommendations. This information will form the starting point for a community discussion about the detailed recommendations and what changes are needed in the effort to implement *VISION 2020*.

CONCLUSION: EVALUATING THE SUCCESS OF THE SUSTAINABLE COMMUNITY INITIATIVE

The overriding goal of Hamilton–Wentworth's *VISION 2020* Sustainable Community Initiative is to:

> integrate the concept of sustainable development into the decision making of individuals, businesses, community groups, and government agencies by building an ethic of sustainability in all of our citizens.

Trying to evaluate progress in relation to this goal is a difficult task. Obviously, from the discussion presented in this report, a number of significant efforts have been made, particularly by Regional Council and staff, to implement the directions of the community vision. To what extent people really understand the idea of sustainability is a different question.

During February and March 1996, the Environmental Health Program at McMaster University assisted the Regional Environment Department in conducting an evaluation of the Sustainable Community Initiative. The purpose was to assess whether the project was reaching the citizens of the community and the level of awareness in the community. The evaluation involved a telephone survey of a randomly selected sample of 250 households in Hamilton–Wentworth. Participants were asked a number of questions designed to determine their general awareness of VISION 2020 and the concept of sustainability. The results of the telephone survey suggest that some success has been made over the last seven years in increasing awareness about sustainable development. Almost one-quarter of the respondents claimed to have at least heard about VISION 2020 or sustainable development. When those people were asked to provide their definition of sustainable development, almost 75% gave a definition that fits with the project's presentation of the concept. Although this study was not as rigorous as it could have been due to financial limitations, it does suggest that somewhere in the range of 10 to 15% of the population (30,000 people) has an understanding of sustainability (Regional Municipality of Hamilton–Wentworth, 1996d). This is a significant improvement from 1993, when an independent study suggested that only about 5% of the population was aware of the concept (Social Planning and Research Council, 1993). Whether people are making changes in their behavior and decision making could not be evaluated because of financial constraints.

Another indicator of whether sustainability is becoming part of the decision-making processes is to review how the local media report on issues of significance. Between 1990 and 1997, the major local newspaper, The Hamilton Spectator, published 75 articles that either are about or make reference to the VISION 2020 initiative. The majority of these articles are reports on specific events or activities that have been sponsored under the banner of VISION 2020, such as the community forums held by the task force, the Children's Sustainability Fair, and the Crazy Commute Day. However, an increasing number (70% since January 1996) of the articles discuss specific issues, such as planning for bicycles, protection of natural areas, or improving water quality, and these issues are being presented in the context of either VISION 2020, sustainable development, or Hamilton–Wentworth's role as a Local Agenda 21 Model Community. The public debate, at least as being presented in the local media, is increasingly occurring within the context of sustainable development and considerations about the long-term future of the community.

There is now a heightened awareness and understanding about sustainability, environmental protection, and the need to find a balance between the economy and environment within the community and in the government bureaucracy. In the latter, there is a breakdown of departmental silos so that cooperation and communication can occur to enhance the sustainable region. This breakdown is also associated with a commitment by regional politicians and management to ensure that sustainability impinges on decision making. There is a tremendous commitment, sense of purpose, consensus, and sense of strategic importance in making the region sustainable. Indeed, this has been recognized nationally and internationally.

In October 1993, the region of Hamilton–Wentworth was selected as the sole Canadian municipality and as one of only 14 communities around the world to serve as a *Local Agenda 21 Model Community*. The International Council for Local Environmental Initiatives (ICLEI) (1996) has in part modeled its guide for local governments on the region's experience. In December 1994, the region received from Environment Canada the *Canadian Award for Environmental Achievement* in the category of leadership by a municipal government. Canada's National Round Table on the Environment and Economy (1994) has profiled Hamilton–Wentworth as an example of how to take the steps toward sustainability at a local level. In fact, in May 1994 the members of the round table came to Hamilton–Wentworth to meet with the people involved in moving the region down the road to sustainability.

Hamilton–Wentworth has set a standard for local sustainable development that is being recognized both nationally and internationally. Over 300 communities and agencies from over 40 countries have requested information about Hamilton–Wentworth's sustainable community activities, while delegations from Argentina, Brazil, China, Gaza, Italy, Jamaica, Mexico, Russia, the Ukraine, Vietnam, Taiwan, the Philippines, and Thailand have visited to learn first-hand about the Hamilton–Wentworth experience.

The region's success has mobilized government, industry, and citizens to address a whole range of environmental and economic concerns. Yet there remain significant barriers. Community awareness is high but requires further attention to move to the next stage of strategic development.

Another significant barrier is people's value set. Establishment of an ethic of sustainability in citizens means the region's simple social marketing activities must compete with the massive marketing budgets of large corporations. Although the municipality can change its way of operation and try to establish itself in a leadership role, it still must respond to the wishes of its citizens. If they are unwilling to accept bicycle lanes, naturalized parks, a more compact urban form, and other elements of the vision, these changes will not occur. On its own, a municipality does not have the resources to create the more fundamental change required in people's attitudes and values.

In fact, the region is part of larger economic and political ideological systems. In Ontario and Canada, the mid-1990s have seen attention directed at deficit reduction, economic growth, and wealth creation with less concern for environmental despoliation or remediation. The national framework for environmental protection and sustainability — the *Green Plan* — was displaced for a voluntarist strategy in 1995. The provincial regulatory regime has been significantly compromised by new more liberal environmental assessment legislation and the withering of the environmental ministry. Ideologically, there is explicit affirmation of individual values and capacities at the expense of the protection of collective and public goods. Yet, there remain glimmers of success and hope — the growth of environmental repair in the industrial fabric of Hamilton–Wentworth, the bureaucratic commitment to sustainability in its decision making, and the citizen awareness of *VISION 2020.* These provide an ethos for practice in the region that will allow, within the limits of what is possible in a local context, Hamilton–Wentworth to remake itself as a sustainable community.

Further, Hamilton–Wentworth provides some practice guidelines for other jurisdictions that wish to adopt its political and community approach to sustainability. There has to be clarity of purpose, often requiring a visioning stage, a commitment on the part of elected officials, flexibility in allocating staff and budgets, and a desire to build consensus, which may be slow and requires patience (see National Round Table, 1993) and which may sometimes break down as different levels of government (the region and its municipalities) have different agendas, and a process to monitor, report, and evaluate progress toward the goals. These characteristics are not unusual; they are found in many U.S. and Canadian cities and communities which desire economic betterment (e.g., Pittsburgh, Cincinnati). But what distinguishes Hamilton–Wentworth from most other places is the tempering of economic benefits by the commitment to sustainability — a difficult but ultimately the only sustainable way.

AUTHORS' NOTE

This chapter draws heavily on a report on the Agenda 21 planning process for ICLEI (Regional Municipality of Hamilton–Wentworth, 1997). Early documentation of the task force (Poland, 1993) was also helpful. The views are those of the authors and do not necessarily reflect the official position of the Regional Municipality of Hamilton–Wentworth.

REFERENCES

Block, P. (1987). *The empowered manager.* San Francisco: Jossey-Bass.

Canada (1991). *The state of Canada's environment.* Ottawa: Ministry of Supply & Services.

Canada Law Book (1996). Dundas (Town) Official Plan Amendment 23. In *Ontario Municipal Board reports,* Vol. 32 (pp. 257–197). Aurora, ON: Author.

Canadian Standards Association (1996). *A guide to public involvement.* Ottawa: Author.

Dear, M., Drake, J., and Reeds, L. (Eds.) (1987). *Steel city — Hamilton and its region.* Toronto: University of Toronto Press.

Eyles, J. and Peace, W. (1990). Signs and symbols in Hamilton. *Geografiska Annaler, 72B,* 73–88.

Eyles, J., Cole, D., and Gibson, B. (1996). *Human health in ecosystem health.* Windsor, ON and Detroit, MI: International Joint Commission.

Health Canada (1992). *A vital link.* Ottawa: Ministry of Supply and Services.

International Council for Local Environmental Initiatives (1996). *The local Agenda 21 planning guide.* Toronto: Author.

Kendrick, M. and Moore, L. (1995). *Reinventing our common future.* Hamilton: EcoGateway.

McKnight, J. (1995). *The careless community.* New York: Basic Books.

National Round Table on Environment and Economy (1993). *Building consensus for a sustainable future.* Ottawa: Author.

National Round Table on Environment and Economy (1994). *Hamilton–Wentworth changes course* (newsletter). Ottawa: Author.

Poland, B. (1993). *A participant-centred evaluation of public participation.* Unpublished master's thesis. Toronto: York University.

Putnam, R. (1993). *Making democracy work.* Princeton, NJ: Princeton University Press.

Regional Municipality of Hamilton–Wentworth (1989). *Directions for the nineties.* Report to Regional Council.

Regional Municipality of Hamilton–Wentworth (1990a). *Task Force on Sustainable Development terms of reference.* Hamilton: Report to Regional Council.

Regional Municipality of Hamilton–Wentworth (1990b). *Task Force on Sustainable Development, Summary report no. 1: Community workshops.* Hamilton: Author

Regional Municipality of Hamilton–Wentworth (1990c). *Task Force on Sustainable Development, Summary report no. 2: Community focus groups.* Hamilton: Author.

Regional Municipality of Hamilton–Wentworth (1990d). *State of the environment, 1990.* Hamilton: Author.

Regional Municipality of Hamilton–Wentworth (1991a). *Task Force on Sustainable Development, Discussion paper series.* Hamilton: Author.

Regional Municipality of Hamilton–Wentworth (1991b). *Task Force on Sustainable Development, Summary report no. 4: Working group final reports.* Hamilton: Author.

Regional Municipality of Hamilton–Wentworth (1991c). *Task Force on Sustainable Development, Summary report no. 3: Community forum, creating the sustainable region.* Hamilton: Author.

Regional Municipality of Hamilton–Wentworth (1992a). *Task Force on Sustainable Development, Summary report no. 5: VISION 2020 community meeting.* Hamilton: Author.

Regional Municipality of Hamilton–Wentworth (1992b). *VISION 2020: The sustainable region.* Hamilton: Author.

Regional Municipality of Hamilton–Wentworth (1992c). *Task Force on Sustainable Development, Implementation Team final reports.* Hamilton: Author.

Regional Municipality of Hamilton–Wentworth (1992d). *Task Force on Sustainable Development, Summary of report no. 6: Community workshop, creating a sustainable region.* Hamilton: Author.

Regional Municipality of Hamilton–Wentworth (1992e). *Hamilton–Wentworth employment trends, 1982 to 1990.* Hamilton: Author.

Regional Municipality of Hamilton–Wentworth (1993a). *Implementing VISION 2020: Directions for creating a sustainable region.* Hamilton: Author.

Regional Municipality of Hamilton–Wentworth (1993b). *Implementing VISION 2020: Detailed strategies and actions for creating a sustainable region.* Hamilton: Author.

Regional Municipality of Hamilton–Wentworth (1994a). *State of the environment update.* Hamilton: Author.

Regional Municipality of Hamilton–Wentworth (1994b). *Official regional plan.* Hamilton: Author.

Regional Municipality of Hamilton–Wentworth (1995). *Young Citizens for a Sustainable Future program.* Working Series Paper No. 4. Hamilton: McMaster University, Environmental Health Programme.

Regional Municipality of Hamilton–Wentworth (1996a). *Sustainable community decision-making guide.* Hamilton: Report to Regional Council.

Regional Municipality of Hamilton–Wentworth (1996b). *Hamilton–Wentworth sustainability indicators. 1995 background report.* Hamilton: Author.

Regional Municipality of Hamilton–Wentworth (1996c). *A guide to pollution prevention for municipalities.* Sarnia, ON: Great Lakes Pollution Prevention Center.

Regional Municipality of Hamilton–Wentworth (1996d). *VISION 2020 evaluation plan 96-036.* Report to Regional Council. Hamilton: Author.

Regional Municipality of Hamilton–Wentworth (1997). *Summary of the model communities program. Report to ICLEI.* Hamilton: Author.

Social Planning and Research Council (1993). *Community monitor.* Hamilton: Author.

Weaver, J. (1982). *Hamilton — A history.* Toronto: Lorimer.

Sustainable Regional Community
Development Cases

The Henry's Fork Watershed Council: Community-Based Participation in Regional Environmental Management

Kirk Johnson

> For too many years, various interests on the Henry's Fork did not listen to each other and instead talked about each other, often through the press....We talked about the law, about man's devastation of the environment, about economics and supporting families and communities, but we didn't communicate to tie these issues together.
>
> —John W. Keys III, 1995

The Henry's Fork Watershed Council is an innovative consensus-based forum for citizens with diverse interests, joined by their common concern to advance the ecological health of the Henry's Fork Basin and the economic sustainability of their communities. Participants include farmers, conservationists, agency and community representatives, elected officials, and others who "reside, recreate, make a living and/or have legal responsibilities" in the 1.7-million-acre basin in eastern Idaho. The council came together as an alternative to the conflict and polarization that had marked resource management debates in the basin for at least two decades and that had grown especially intense in the early 1990s. The council has come to be seen as a significant and leading experiment in ecosystem management that integrates the needs and desires of people. The council provides an example and specific lessons on negotiating the complex relationships among national, state, and local institutions and their constituents.

THE HENRY'S FORK BASIN

The Henry's Fork river is a tributary of the Snake River, one of the longest in North America.[1] The 1.7-million-acre Henry's Fork Basin encompasses over

3,000 miles of rivers, streams, and irrigation canals in eastern Idaho and western Wyoming, including the southwest corner of Yellowstone National Park. The basin supports numerous healthy fish and wildlife populations, as well as several threatened and endangered species, including bald eagles, whooping cranes, grizzly bears, and gray wolves. It provides high-quality recreational experiences for Idaho residents and visitors. The population of the basin is 40,000, spread over three Idaho counties and one Wyoming county. Agriculture is the major industry. More than 235,000 acres of farmland are irrigated from surface or groundwater sources; potatoes and grains are the primary crops. Other important economic activities include recreation and tourism services, government programs, and timber production.

During the past two decades, increasing demands were made on the Henry's Fork to meet irrigation, hydropower, and in-stream flow needs for fisheries and recreation. These demands brought increasing conflict among groups with an interest in the river. In 1992, two separate environmental incidents inflamed passions on all sides: an accident during construction of the Marysville Hydroelectric Project on the Fall River and planned drawdown of the Island Park Reservoir, which dumped over 50,000 tons of sediment into the Henry's Fork through Harriman State Park. The following year, the Idaho legislature passed a compromise Henry's Fork Basin Plan, which restricted new development on 195 miles of the river and its tributaries and called for improvements in water quality, fish and wildlife protection, and water conservation.

FORMATION AND STRUCTURE OF THE WATERSHED COUNCIL

In June 1993, a community meeting was held, with many agencies in attendance, to discuss the sediment spill of the previous year. Attendance was large and many impassioned speeches were made. Agreement was reached on only one item: the watershed community had to find a better way to coordinate agency actions and involve the public. A subcommittee was appointed to develop the conceptual structure for what would become the Watershed Council. In August 1993, the subcommittee proposed that a watershed council be formed with three primary interests equally represented: government agencies, scientists, and the public. Thus, the Watershed Council was formed.

A significant moment occurred during the August meeting when it came time to identify facilitators of the council. Janice Brown, the assertive director of the Henry's Fork Foundation,[2] volunteered her services. "The room was silent. We didn't think the Council would work with only the leadership of the Henry's Fork Foundation, because some would label it as an 'environmental group,'" recounted one participant. To the group's surprise, Dale Swensen, the more reserved director of the Fremont–Madison Irrigation District, which had clashed many times with the Henry's Fork Foundation, volunteered to co-facilitate. How could an environmental group and an irrigation district work together? "The collaboration between Jan and Dale has been exemplary and has held the Council together. Both have put much thought into the Council and its impact on the community. Most importantly, the Council has not been classified as representing any one particular interest. People of all viewpoints feel welcome the first time they attend a Council meeting" (Keys, 1994, pp. 1–2). "Jan and Dale bring out the best in each other," said one observer.

Brown, Swensen, and others from their organizations selected as members of a facilitation team for the council received initial training in meeting facilitation from the Foundation for Community Encouragement, a non-profit educational organization devoted to community building.[3] The council's early months were marked by a series of trust-building activities that established the foundation for more difficult deliberations down the road. These included developing the council's legislative charter and mission statement by consensus.

Charter, Mission, and Goals

The Idaho legislature approved the council's consensus-based charter and mission statement in 1994. The broad goals set out in the mission statement are:

- To serve as a grass-roots community forum which uses a non-adversarial, consensus-based approach to problem solving
- To better appreciate the complex watershed relationships in the basin, to restore and enhance watershed resources where needed, and to maintain a sustainable watershed resource base for future generations
- To respectfully cooperate and coordinate with one another and abide by federal, state, and local laws and regulations

Four related major duties for the council, identified in the charter, are to (1) cooperate in resource studies and planning that transcend jurisdictional boundaries; (2) review, critique, and prioritize proposed watershed projects; (3) identify and coordinate funding for research, planning, and implementation and long-term monitoring programs; and (4) serve as an educational resource for the state legislature and the general public on the council's progress.

The Council's Consensus Decision-Making Process

Drawing on Quaker tradition, Watershed Council meetings begin and end with all participants seated in a single large circle. These community-building sessions start with several minutes of silence, after which participants are encouraged to "speak from their hearts" about issues of importance to them and the council. Carefully arranged ground rules help to maintain a safe atmosphere for the meetings, where differences can be aired in a non-hostile way. Chief among these are "I" statements, so as not to assign motives to others and to avoid personal attacks (see the appendix at the end of this chapter for a list of the rules used for council meetings).

According to participants, the council fills a need in the community for a forum where people can talk about problems or issues in a civil manner. It builds a sense of trust among participants and a growing understanding of each other's perspectives. Meetings have demonstrated that the various interests represented have more in common than one might think. Explains co-facilitator Janice Brown:

> Many folks fear consensus for it mistakenly implies a drawn out process and a yielding of fundamental beliefs or philosophies. In our experience, the time involved in building consensus has been well used, and it has not resulted in win/ lose solutions or compromising one's values. Rather, after being heard and considering all sides, people who know consensus is required will fashion creative solutions acceptable to all, and retain their dignity and mutual respect in the process (personal communication, February 23, 1995).

Meetings of the council occur eight times per year (less frequently during the summer due to farm work demands). The meetings are open to all who wish to attend; participants are encouraged to attend as individuals, although they may be official representatives of an interest or an organization. Attendance averages around 50 participants and includes a diversity of people — a high number of agency representatives and citizens representing farming and conservation interests and a smaller number of community and elected officials, along with researchers and observers from other parts of Idaho and the West. The council does not have a board of directors, but is led by the facilitation team and co-facilitators Brown and Swensen.

Meetings all begin with the council reflecting on the meeting's objectives as a whole body, followed by educational or project presentations. In the afternoon, the council breaks into three component groups for detailed discussion and problem solving. The three groups include: (1) an agency round table consisting of representatives from state, federal, and local agencies with rights or responsibilities in the basin, as well as representatives of the Native American Shoshone Bannock tribes; (2) a technical team composed of scientists and technicians from universities, government agencies, and the private sector; and (3) a citizens group which includes members of the public with commodity, conservation, and community development interests.

Funding

The council is administered jointly by the Fremont–Madison Irrigation District and the Henry's Fork Foundation, which are chartered to attend to the council's administrative and logistical needs on a voluntary basis. The Henry's Fork Watershed Fund was established by the state of Idaho to help fund projects in the basin and to defray the council's out-of-pocket expenses. The U.S. Bureau of Reclamation contributed $20,000, initially. Another $150,000 was received from the Marysville Hydroelectric Partners, as settlement for the Marysville incident from June 1992. The council expects to disburse between $20,000 to

$25,000 per year, through fiscal year 1999, to projects in the basin. In addition, the council seeks grant money and private contributions.

INITIATIVES AND ACCOMPLISHMENTS

Watershed Integrity Review and Evaluation (WIRE) Criteria

One of the council's first tasks was development of its WIRE criteria. More than 80 different ideas for watershed health and vitality were consolidated into 10 primary criteria for ensuring the integrity of the Henry's Fork Basin. The criteria, formatted as a checklist, are used to evaluate the merits of projects or programs brought before the council by agencies or other council members. These criteria reflect the council's mission and goals:

1. *Watershed Perspective* — Does the project employ or reflect a total watershed perspective?
2. *Credibility* — Is the project based on credible research or scientific data?
3. *Problem and Solution* — Does the project clearly identify the resource problems and propose workable solutions that consider the relevant resources?
4. *Water Supply* — Does the project demonstrate an understanding of water supply?
5. *Project Management* — Does project management employ accepted or innovative practices, set realistic time frames for their implementation, and employ an effective monitoring plan?
6. *Sustainability* — Does the project emphasize sustainable ecosystems?
7. *Social and Cultural* — Does the project sufficiently address the watershed's social and cultural concerns?
8. *Economy* — Does the project promote economic diversity within the watershed and help sustain a healthy economic base?
9. *Cooperation and Coordination* — Does the project maximize cooperation among all parties and demonstrate sufficient coordination among appropriate groups or agencies?
10. *Legality* — Is the project lawful and respectful of agencies' legal responsibilities?

"WIREing" projects has become one of the important functions of council meetings. Projects that gain council approval may be eligible for some of the Watershed Funds the council has available for projects or backers may use the endorsement to seek funding from other public or private sources. So far, the council has assessed over 20 projects presented to it. Most of those projects were approved outright (some with conditions attached) and three were approved after suggested improvements were made.

Public Education and Participation Initiatives

To promote public understanding of issues facing the watershed, the council has sponsored a number of educational events. Examples include:

- A forum on Columbia River salmon recovery, in July 1994. Salmon recovery is a major political issue in the entire northwest region of the United States. Representatives from the National Marine Fisheries Service and the Northwest Power Planning Council and various other interest groups attended. Following the forum, the council used the WIRE process to develop and issue *A Consensus Position on the Recovery of Snake River Salmon and Steelhead.*
- An annual State of the Watershed Conference, which provides participants an opportunity to learn about the numerous research and management projects occurring in the basin. The October 1996 conference drew more than 100 attendees.
- A seminar entitled *Water Rights 101,* in April 1995, which provided participants with a brief overview of Idaho water law and how the "Prior Appropriations" doctrine,[4] regarding the administration of water rights, applies in the state.
- A forum, in June 1995, which addressed the contentious issue of road access to the Targhee National Forest. The forum heard from representatives of the U.S. Forest Service, the U.S. Fish and Wildlife Service, and various other interest groups. Another educational session was held in March 1996 to provide information to the public on the draft plan. A year later, the final plan was first announced to those attending the April 1997 council meeting.

One of the council's current successful projects is coordinating a major land and water restoration project on Sheridan Creek. A number of federal and state agencies and community groups are contributing to the project in various ways. This project illustrates how the council encourages the sharing of information, knowledge, and resources (financial and in kind) present within the group. It helps to provide a better scientific understanding of how the watershed works. It provides a forum for the coordination of research projects and the presentation of results, promoting watershed literacy among participants.

Facilitating Community Relationships with Government Agencies

The council is developing an innovative role for itself as a facilitator of improved dialogue between the public and representatives of local, state, and federal agencies which manage lands and resources. Government agency representatives are generally complimentary of the council. The council provides them with a context for meeting with groups and people they might not otherwise encounter — including other agency officials — and to learn about activities developing in the community. Thus, the council plays an important coordination role. Projects submitted to the council through the WIRE process reveal areas of possible duplication or omission, which then can be addressed, appropriately. In a basin as large as the Henry's Fork, no one organization can embrace or fully apply the concept of ecosystem management. Thus, the council's coordination function is helpful on issues that transcend the understanding or jurisdiction of any one agency.

However, relationships with government agencies remain complex and delicate. Agencies may feel threatened by the council, although they maintain all legal authority. Moreover, the Federal Advisory Committees Act (FACA) governs how federal agencies are required to behave in involving the public in decision-making processes, and it is designed to prevent any one group from wielding undue influence. Agencies such as the U.S. Forest Service have interpreted FACA to strictly limit their participation in forums unless they themselves charter the groups and appoint representatives. Nonetheless, certain features of the Henry's Fork Watershed Council's organizational makeup have enabled it to avoid conflict with FACA: its facilitation by two non-profit-sector organizations (as opposed to a federal agency), its open participation policy, its lack of an exclusive board of directors, and the fact that it has no formal or legal influence over management decisions of participating federal agencies. Other communities considering formation of voluntary watershed councils would do well to consider carefully their organizational relationship to government agencies. Policy reforms are likely to accelerate in years to come, creating opportunities for such voluntary institutions to play an increasingly influential role in land and resource management.

COMMUNITY BUILDING

More than any specific action or initiative, the council's greatest accomplishments have been in encouraging former adversaries to work together, in a non-hostile setting, to develop common goals and objectives for the sustainability of the Henry's Fork Basin. Many council participants and observers did not think the council would work at all, let alone survive and flourish for several years with a growing list of accomplishments. Participants who were interviewed expressed high praise for the council. Positive feelings about the group, its participants, and process are tangible during meetings. One of the most common sentiments is that the council maintains a "safe" and "friendly" atmosphere for discussing potentially contentious issues. It fills a need in the community for a forum where people can talk about problems or issues in a civil manner. It is building trust and understanding of differing perspectives.

The participants have learned that they have more in common than they thought. For example, farmers and conservationists each thought they were the only ones who wanted good science. Both groups have discovered it is a mutual desire. Council meetings represent a group education process in which participants are both students and teachers. The process produces crosscurrents of opinion and perspective, providing a broader understanding of problems and possible solutions than any one individual or group could generate. This is the promise and the benefit of the consensus-based process at the core of the council's work. This process also helps participants stay ahead of problems — providing an opportunity for input on projects before they are launched, reducing costly and damaging mistakes, and reducing the need for corrective solutions imposed from outside. Most of all, the council's process is helping residents of the Henry's Fork Basin to expand their capacity for self-governance — their capacity to discuss, evaluate, and resolve issues and conflicts on their own. Ronnie Ard, farmer and member of the facilitation team, says:

> This is the best group I'm involved with. It gets a lot done. It's not an issue of power. No one is seeking power. We want to solve problems. You can say what you feel without getting jumped on. It's changed attitudes about people in the community. The differences are not so great between environmentalists and farmers. A big wall had been built. First we helped people peek over, and now we're tearing it down.
>
> People want to feel involved in the process. You don't feel that with government processes where you can say what you think but you don't know whether your comment is heard, or how decisions are made. Here, people feel ownership. They feel their view is considered. They are part of the process. People live and work here together. They want to protect the area, and to get along with each other (personal communication, April 19, 1995).

CHALLENGES FACING THE COUNCIL

As noted earlier, the council needs to work to ensure feasible and consistent working relationships with government agencies that have jurisdictional authority in the region. Such regular cultivation of the relationships can help defuse some of the anti-government sentiment, which is fairly widespread in the West. And, it can play a constructive role in helping agencies develop more effective ways to work with the public on difficult issues.

A common sentiment among participants is that the council has yet to be truly tested. There is little doubt that such a major test, such as another drought or a serious proposal to rebuild the Teton Dam,[5] will come. While the council has confronted a number of weighty and contentious issues, these have not so far pitted the two facilitator groups (the Henry's Fork Foundation and the Fremont–Madison Irrigation District) against one another. It remains to be seen if the council has the institutional strength to survive such a test. Moreover, the council must strike a balance between seeking out contentious issues, which could jeopardize its working relationships and its unique role in the basin, and becoming too timid, which could cause it to lose participation and relevance to real problem solving. It was the initial development of this plan six years ago that contributed to the intense polarization in the community preceding formation of the Watershed Council.

The council needs to continue to work at expanding participation by a variety of groups, such as farmers who are not members of the Fremont–Madison Irrigation District, local elected government officials (i.e., in comparison to agency bureaucratic representatives), recreational guides and outfitters, and community leaders. Broader involvement will make consensus more difficult, but the council would thereby be more truly representative of all residents, better able to address a broader range of community issues, and more influential as an organization. Some council participants also have great interest in expanding participation by schools, to perhaps influence a new generation of community leaders in collaborative processes.

The council has not directly tackled the intersection between resource management and economic development. One area it could explore is the use of consumer demand to modify resource management practices. Numerous efforts are developing in the Northwest (United States) to examine incentive-based, market-driven approaches to achieving desired environmental goals. Examples include certification of sustainably grown and harvested agricultural crops, seafood, and forest products. The hope and promise is that demand for such products will ultimately create a powerful economic incentive for landowners to adopt more environmentally benign resource management practices. Such an approach would be a positive, non-adversarial way for the council to address sustainable agriculture practices, including the reduced use of pesticides and chemical fertilizers.[6]

THE COUNCIL'S SIGNIFICANCE FOR RESOURCE MANAGEMENT IN THE AMERICAN WEST

The council is one of a group of experiments emerging in the American West for new, more responsive, and potentially more effective ways to manage and maintain healthy ecosystems while integrating the needs and desires of people and communities within the region. These efforts have arisen for a variety of reasons, including pressure to reform ways in which public agencies interact with the public, a growing concern over the long-term effects of traditional interest group advocacy (i.e., adversarial) tactics, and a movement among resource managers toward more complex approaches to ecosystem or watershed management.

Don Snow, executive director of the Northern Lights Research and Education Institute in Missoula, Montana, describes the void that groups, such as the Henry's Fork Watershed Council, are helping to fill:

> The public lands of the West stand at the beginning of a massive power shift — a shift away from the centralized management of natural resources and toward community-level involvement in major land management decisions, beginning with the federal lands...
>
> ...[T]he West is groping for new institutions of governance to repair and in some instances replace the region's time-honored reliance on federal agents to make so many key decisions about resource allocation, the protection and conservation of lands and waters, the direction of local economies, pollution control, and the critical ecological relationships between private and public lands. The growing awareness of this trend is creating new and very interesting opportunities for the processes of collaborative problem-solving, for these contain the seeds for creating new forms of governance appropriate to this rapidly changing region (Snow, 1995, pp. 9–10).

Formation of the Henry's Fork Watershed Council was triggered by two accidents, but members focused, instead, on the well-being of the entire watershed. An uncertain political climate also pushed the various interests together, as none could count on advancing its long-term interests through the courts, legislature, or Congress. Constructive leadership by a few key individuals helped turn the situation in a positive direction and directed the group to find purpose out of chaos. Formal training of the leadership in consensus decision making and leadership skills also fostered the council's success. The council's open participation rules and its organizational design have helped avert polarizing conflict with groups such as national (non-local) environmental organizations and the federal agencies mentioned above. The council's design provides participants a common frame of reference and facilitates an emphasis on community building. This is most likely to succeed where participants have a strong commitment to or love for the community, watershed, or region that counteracts interest group differences. The Henry's Fork Watershed Council is helping to chart a new path for resource management, local governance, and relationships between public agencies and the public they serve.

AUTHOR'S NOTE

This chapter is based on 1995 and 1996 evaluation research studies of the Henry's Fork Watershed Council conducted under the organizational aegis of the Northwest Policy Center, Graduate School of Public Affairs, University of Washington, Seattle. The research was requested by the Henry's Fork Foundation with the support of the Fremont–Madison Irrigation District and funded, in part, by the National Fish and Wildlife Foundation.

REFERENCES

Keys, J.W. III. (1994). Observations on the first year of the Henry's Fork Watershed Council. In R. Van Kirk (Ed.), *Proceedings of the First Annual State of the Watershed Conference.* Rexburg, ID: Henry's Fork Foundation, November 21.

Reisner, M. (1993). *Cadillac desert: The American West and its disappearing water.* New York: Penguin Books.

Snow, D. (1995). New governance? Or return to old apathy. *Northern Lights,* Spring, 9–10.

APPENDIX*

Some Rules of Community Building

One of the most important elements of community is authentic, effective communication. The following communication skills are essential:

- Use "I" statements when speaking.
- To communicate effectively, speak personally and specifically, rather than generally and abstractly.
- Listen to your inner voice. Become aware of when you are moved to speak and when you are not moved to speak.
- Listen carefully and with respect to what another person is telling you. Do not formulate your response while someone is speaking, but wait until the other has completely finished.
- Be aware of your own barriers, such as prejudices, expectations, ideologies, judgments, or a need to control, which are obstacles to community.
- Be willing to share your own woundedness. This way, you invite others to be vulnerable with you.
- Sharing brokenness as well as heroism is an essential part of maintaining community. Both the darkness and the light can be expressed.
- Understand the value of silence in communication. Be comfortable with silence, your own and that of others.

Some Principles of Community

- Community is inclusive. Individual differences are celebrated. Soft individualism, rather than rugged, can flourish.
- Community is realistic and multidimensional. Each member is free to experience his or her own facet of reality.
- Community facilitates healing once its members stop trying to heal or fix one another.
- Community is reflective, contemplative, and introspective.

* Source: Foundation for Community Encouragement.

- A community's members can fight gracefully.
- A community is a group of all leaders who share equal responsibility for and commitment to maintaining its spirit.
- A community is a highly effective working group.
- A community is the ideal consensual decision-making body.
- In community, a wide range of gifts and talents is celebrated.

ENDNOTES

1. The Henry's Fork is a tributary of the Snake River (1,038 miles in length), which in turn drains into the mighty Columbia, the fourth largest river in North America (Reisner, 1993, p. 155). The Columbia empties into the Pacific Ocean at Astoria, Oregon.
2. The Henry's Fork Foundation is a non-profit organization dedicated to understanding, restoring, and protecting the unique fishery, wildlife, and aesthetic qualities of the Henry's Fork of the Snake River (mission statement).
3. The Foundation for Community Encouragement can be reached at P.O. Box 17210, Seattle, WA 98107 (telephone: 206-784-9000).
4. The "Prior Appropriations" doctrine states, with respect to water use rights, that those who are "first in time" are "first in right" and that water must be put to "beneficial use."
5. The Teton Dam, built on the Teton River in 1975, was a major engineering disaster. When the dam broke in 1976, damages were estimated at $2 billion, and the topsoil was stripped from tens of thousands of acres (Reisner, 1993, pp. 407–408).
6. However, the Henry's Fork Basin has had few problems with leaching of nitrates or chemicals (Janice Brown, personal communication, May 5, 1997).

CHAPTER 10

The Willapa Alliance: The Role of a Voluntary Organization in Fostering Regional Action for Sustainability

Marie D. Hoff

> In Willapa, the land, the waters and the lives of people are inseparable
>
> —Wolf, 1993, p. 5

Willapa Bay is located in the southwest corner of Washington State, separated from the Pacific Ocean to the west by a narrow 28-mile-long peninsula. East of the bay are the Willapa Hills, covered by towering evergreens of the temperate rain forest of the North American coastal range. To the south, another narrow strip of land separates the bay from the mouth of the mighty Columbia River. Numerous smaller rivers, flowing down from the forested mountains, send fresh water into the bay, while the salty Pacific tidewater washes into the bay from the opposite direction twice every day. Willapa Bay and its encompassing ecosystem (some 680,000 acres) is a place of exceptional beauty, and it is exceptional in several other respects. Willapa Bay may be the cleanest estuary in the continental United States, and it is one of the few relatively unspoiled estuaries in the world. Native Americans introduced the early white explorers and settlers to the succulent and plentiful oysters of the bay in the nineteenth century. Today, Willapa Bay remains one of the most productive oyster-growing areas in the world. The coastal rain forest receives anywhere from 85 to 200 inches of rain per year. Trees grow faster here than almost anywhere in the United States. Three species of salmon — chinook, coho, and chum — spawn in the numerous rivers and streams which flow down from the mountains into Willapa Bay. Elk, deer, bears, owls, eagles, waterfowl, shorebirds, and other North American wildlife grace the forests and waters. Native cranberries, harvested by the Chinook Indians, have been replaced by introduced vines, which produce the ecosystem's most valuable agricultural crop. The Long Beach peninsula, which

1-57444-129-9/98/$0.00+$.50
© 1998 by CRC Press LLC

protects the bay from the Pacific Ocean, boasts a 28-mile-long sandy beach, which attracts thousands of tourists every year. However, the relative remoteness of the area — several hours drive from the major Northwest cities of Seattle and Portland — has helped slow population growth. The permanent population is about 20,000 residents. "Since the first permanent white settlers arrived, Willapa's principal livelihoods — oyster harvesting, logging, fishing and tourism — have remained the same" (Wolf, 1993, p. 18).

ENVIRONMENTAL AND ECONOMIC RESOURCES AND THREATS

But the Willapa ecosystem has changed significantly since the first white explorers encountered the native Chinook, Chehalis, and Kwalhioqua peoples. Foreign species, such as the spartina grass, which threatens the intertidal zone of the bay, have been introduced; timber-harvesting methods have contributed to habitat loss for salmon; salmon decline, in turn, may have contributed to the increase in burrowing shrimp in the bay, which degrade the productive oyster beds; and less than 5% of the old-growth forest remains. Although human population increase (in the form of tourists and retirees) is at a modest pace, it also contributes to changes in the Willapa watershed and in the local socioeconomic conditions of the communities surrounding Willapa Bay.

In the late 1970s and 1980s, the regional economic system also experienced significant change, in the form of major declines in the forestry and fishing industries. Pacific County (the political jurisdiction which covers most of the Willapa ecosystem) is one of the poorest counties in Washington State; young people who traditionally would have looked forward to good-paying jobs in the extractive industries found the range and level of employment opportunities in their local community severely constrained.

Threatened Salmon

Every natural ecosystem is comprised of a diversity of species which support one another in a functional way. For human cultures, certain plant or animal species tend to achieve central economic and totemic significance. In the Pacific Northwest of North America, the salmon undoubtedly occupied this position for the indigenous peoples of the region. Work, food, religion, and a rich artistic heritage developed around the salmon and its complex, mysterious life cycle and migration between freshwater streams and the salty ocean. Historically, the abundance of the salmon was the stuff of legend — in some places, people said, they were so abundant one could walk across a river on their backs (Wilkinson, 1992). From an estimated stock of 16 million a century ago, the Columbia River Basin salmon stock is now down to 2.5 million or less (Golec, 1996, p. 9). In the food chain, salmon support both ocean marine life and many species of animals and birds, especially eagles and osprey during the freshwater phase of their migrations. Even after death, their carcasses provide a natural fertilizer.

The threatened salmon serve as a key indicator of the general degradation of the regional environment and the unsustainability of numerous human practices, such as logging and the construction of the numerous hydroelectric dams along the Columbia (the fourth largest river in North America) and its many tributaries. The local Willapa ecosystem has not been affected by dams; the local factors of decline are artificial salmon propagation practices and stream degradation from logging. People of the Northwest retain a deep fascination with and concern for the salmon; salmon decline has cost the region dearly: in loss of economic and cultural diversity (cultural health), in dollars and jobs lost, and in fierce battles among the interests involved (Native tribal and non-Native fishers, commercial and sport fishers, Canadian and American fleets, and environmentalists against so-called economic interests).

Oysters: The Canary in the Waters

As noted above, Willapa Bay is one of the richest oyster-producing areas in the world. One-sixth of all the oysters produced in the United States are grown here. The native Olympia oyster is long gone, except for a few remnant stocks. The hatchery cultivation of Japanese oysters has enabled the region to maintain oystering as a major economic enterprise. All the parties interviewed for this study credited the oyster growers' key role in sustaining the cleanliness and health of Willapa Bay. Oysters filter the water and thus provide immediate feedback on the condition of the bay. To date, just one confined incidence of *E. coli* bacterial outbreak has occurred in the oyster beds, due to deficient septic tanks, which have since been replaced. This is one of only two instances in which the state of Washington recertified an area for safety after discovery of *E. coli* bacteria — lending evidence of the locality's firm determination to maintain the health of Willapa Bay. Potential environmental threats to the oysters are numerous: bacterial infection from wildlife, cattle farms, and human waste and burrowing mud and ghost shrimp which destroy the mud flats required by the oysters (as well as by crab and native eel grass which help maintain the bay's health). The shrimp, too, are native to the bay, but may have burgeoned out of control due to decline of salmon and sturgeon which feed on them. An additional threat is spartina grass, a non-native grass, which also contributes to destruction of the mud flats. Efforts to control the shrimp and invasive spartina grass with chemicals have been controversial in the local community, and difficulty in agreeing on a solution has lost valuable time in controlling these species (Hunt, 1995a, p. 7; Herb Whitish, personal communication, October 11, 1996). Threats to the delicate ecological balance in the bay itself demonstrate how environmental threats converge and interact to magnify problems: invasions of foreign species, upset in the balance among diverse and interdependent species, and the potential for pollution from human activity. Population growth in the area, especially on the popular Long Beach peninsula, has made sewage disposal an important political issue (Ann Saari, personal communication, July 16, 1996).

Logging and Forestry

To the east of Willapa Bay, the flanks of the Willapa Hills once held some of the oldest and largest of trees in the world — Douglas fir, western hemlock, western red cedar, and Sitka spruce. For over 100 years, logging was the economic mainstay for residents east of the bay. The first sawmill opened in 1858 (Wolf, 1993, p. 23). Depletion of the old-growth stock, environmental concerns, and government regulation of environmental practices in the industry, as well as changes in the global market, led to major decline in logging and sawmill jobs during the 1980s, and local communities, such as South Bend and Raymond east of the bay, went into major economic decline. The Weyerhaeuser Company, which owns two-thirds of the forest land, has pioneered the practice of "sustained yield" (Wolf, 1993, p. 27) forestry, in which trees are replanted and harvested every 50 to 60 years. The stands of trees are depressingly sterile, from an aesthetic viewpoint.[1] Moreover, earlier logging practices, such as clear-cutting large areas and cutting too close to streams, as well as logging roads, contributed in a major way to the decline of wild salmon, through the soil erosion, sedimentation, and warming of spawning beds which resulted from loss of tree cover. The Weyerhaeuser Company also has contributed funds and support in recent years to an expensive effort to restore salmon habitat through stream rejuvenation.

Other Industries and Socioeconomic Challenges

Cranberries are the region's largest agricultural crop; their need for clean water and their economic value have given growers the incentive to experiment with integrated pest management methods and to control the use of chemical pesticides (Wolf, 1993, p. 42).

On summer weekends, 40,000 tourists, seeking ocean beaches, nature experiences, and numerous other tourist activities and amenities, temporarily explode the population of the region. While tourists bring welcome dollars to retail businesses and services, tourism challenges the delicate ecological and economic balance of the local community. Enchanted by the beauty of the place, some visitors also choose to relocate to this beautiful area for work or retirement. Population growth is not a major problem per se in the Willapa Bay region, but it presents a certain degree of challenge to the local community, such as increased need for social and health services for retirees. On the other hand, new residents may bring new money and possibly new perspectives and ideas into a community.

Pacific County has an unemployment rate around 12% and the rate of poverty is around 17%, higher than the average rates for Washington State. Out-migration of young people, in search of jobs and better salaries, is a concern for the community, according to participants in this study. Health indicators for the community need refinement, and there is an unresolved controversy in the community between some white and Native American residents over health conditions and access to healthcare resources for Native Americans.

With the sharp economic declines in two major extractive industries (forestry and fishing) in the Pacific Northwest, many people began to see more clearly the intimate linkages between economic concerns and environmental concerns. Increased awareness of these linkages, coupled with major problems in both industries, has also contributed to their increased political salience and explosiveness (Dietrich, 1992). Interviews with members of The Willapa Alliance and with otner local leaders revealed that the research and regulatory role of state and federal government officials is a highly charged concern among local residents. The state of Washington has mandated that all counties develop formal land use and growth management plans; other state-level agencies such as the Departments of Fish and Wildlife, Ecology, and Natural Resources have a direct regulatory mandate over various aspects of resource management, as well as over pollution and waste control in all counties in the state. Moreover, the state has a tax and revenue stake in all counties. In Pacific County, both privately owned timberland and state-owned land yield revenue to the state. Various federal agencies have reasons to pursue an active involvement in the county. For example, the Environmental Protection Agency, which has an interest in water quality, has funded local salmon walk (volunteer stream cleanup) programs, the U.S. Army Corps of Engineers has regulatory authority over rivers and wetlands, while the U.S. Geological Survey monitors stream flow. These brief examples indicate the complexity and division of governmental responsibility for environmental monitoring.

The Willapa Bay ecosystem presents a picture of paradoxical environmental and economic conditions: a region of exceptional beauty and natural bounty, yet facing serious environmental threats and struggling to redevelop an economy which will reward its local residents with a reasonable quantity of material wealth and a good quality of life. In this unusual combination of circumstances, The Willapa Alliance is emerging as an organizational leader to assist and enable the residents of this ecosystem to preserve their rich natural heritage while also improving the socioeconomic circumstances of the people who dwell there — Native and white Americans and a growing number of recent immigrants from Mexico and Asia.

FORMATION OF THE WILLAPA ALLIANCE

In the 1980s, when Spencer Beebe, a native of the state of Oregon, was working to save tropical rain forests and consulting internationally on principles of sustainable development, he realized that the credibility of North Americans was compromised by the sorry state of the temperate rain forest in his own home region. Ecotrust (an Oregon-based environmental non-profit organization which Beebe had helped start) and the Washington State chapter of The Nature Conservancy (an international environmental organization) decided to cooperatively investigate the potential for demonstrating the principles of ecologically sound economic development in the Pacific Northwest. The two organizations jointly began a series of informal discussions with some environmentally minded business people, such as the local oyster growers, and other conservation-minded local leaders in the communities surrounding Willapa Bay. They found that they agreed on a basic principle of environmental protection — people take more interest in preserving their environment when they see and experience a sound economy based on that environment (Hollander, 1995, p. 8). After about two years of informal discussions and planning, The Willapa Alliance was started with the fund-raising support and advisory assistance of Ecotrust and The Nature Conservancy and with the participation of local organizations and individuals who shared this vision of fostering ecologically sustainable economic development for the Willapa watershed. "Phase I of our Plan consisted of initial conceptualization, forming partnerships...preliminary research and design work, incorporation, the development of a strategic plan, and leadership recruitment" (1995 Annual Report, p. 2). This phase of development was completed in early 1994. As will be evaluated later, the organization initially faced some local suspicion of its origins and purposes, as more conservative members of the local community wondered if the funding support and advisory role of non-local environmental organizations would lead to outsiders coming in to tell them how to run their local community.

Mission and Programmatic Strategy

Group work theorists playfully suggest that the first three stages of group formation are "forming, storming, and norming." Apparently the founding members of The Willapa Alliance Board of Directors experienced these stages. According to the executive director, meetings during the first year or more of their existence included a lot of "dreaming, scheming, and screaming, followed by steaming — that is, moving forward" (Dan'l Markham, personal communication, July 15, 1996). Eventually the founding members developed a three-phase long-range plan to accomplish the following mission:

> The mission of The Willapa Alliance is to enhance the diversity, productivity, and health of Willapa's unique environment, to promote sustainable economic development, and to expand the choices available to the people who live here (1995 Annual Report, p. 1).

Alliance members voice a strong commitment to three key principles for action which inform all their activities. These include: (1) the generation of sound scientific knowledge of the issues involved as the necessary foundation for decisions and actions; (2) the promotion of civic dialogue and debate about environmental, economic, and social issues facing the community; and (3) the promotion of partnerships among all the community's stakeholders to work creatively "to ensure the long-term well-being of Willapa's communities, lands, waters, and economy" (1995 Annual Report, p. 1).

The Alliance achieved formal incorporation as a not-for-profit civic organization by November 1992; many of those leaders involved in the formation stages remained as board members, although several original participants resigned for a variety of reasons. Like many fledgling organizations, the Alliance experienced threats to its viability. The organization had to work hard to overcome the insinuation, by some local opponents, that it was a tool of outside environmental interests. The first staff director proved unsuitable due to lack of experience in the non-profit sector. According to various board members, the hiring of Dan'l Markham as executive director was a key positive move which stabilized the administrative operations and which helped overcome the accusations of outside control. Mr. Markham's family roots in the local community are deep, as had been his previous public presence in the local community as an elected county commissioner, as a clergyman, and as an active citizen. By the end of 1994, the Alliance had achieved financial independence and had made progress in the development of its four identified program areas, which are Natural Resource Management, Science and Information, Public Education and Involvement, and Conservation-Based Development.

Natural Resource Management

In consideration of the central economic and cultural significance of the salmon, The Willapa Alliance chose to focus one of its first natural resource management efforts on salmon restoration (and other fisheries) through development of a plan entitled *Willapa Fisheries Recovery Strategy* (WFRS). Proceeding on its principles of scientific knowledge and inclusion of stakeholders, the Alliance has generated three natural resource management projects: (1) a study, followed by public discussion, led to agreement to develop the WFRS (other communities in the northwest region are using it as a model to develop similar strategies); (2) *The Willapa River Watershed Restoration Partnership Program,* also a model program, involved about one million dollars worth of work on restoration of salmon habitat and involved the cooperation of a remarkably broad range of stakeholders, including the Weyerhaeuser Company, the (state-level) Department of Natural Resources, local farmers, and local conservation groups (1995 Annual Report, p. 3); and (3) to respond to local controversy over how to control the noxious weed spartina, the Alliance provided leadership to form the Spartina Coordinating Action Group. During the many years of indecision, the invasive weed multiplied dangerously, but the Alliance has used science, science education, and community dialogue to generate supportive legislation, funding, and a methodology to work at controlling the weed.[2]

Science and Information

One commitment which unites the members of The Willapa Alliance is their belief that good science, and public understanding of this science, is an essential foundation from which to build support for the sustainable development goals of the organization. Without a good scientific basis, the organization could make poor social or environmental decisions; moreover, the general public is more likely to support sound decisions if the scientific knowledge about the community is widely disseminated and known. In its short organizational life, the Alliance has generated a notable range of scientific studies and some innovative ideas to share this knowledge among scientists and the general public. For example, the increasing use of computerized Geographic Information Systems (GIS) encouraged the Alliance to participate in development of a GIS database for the Willapa Bay watershed. The solid information which has been built on salmon restoration, to name just one component of the GIS, has helped resolve conflicts and build consensus on highly charged natural resource issues. The Willapa GIS was also used to develop an intriguing public education tool in the form of a multimedia computer disk (CD-ROM), entitled *Understanding Willapa*.[3] This computerized informational tool is available for use in public schools and libraries, as well as for use by individuals. It "...uses salmon as a common thread to weave together the culture, the resources and concerns of the Willapa ecosystem" (1995 Annual Report, p. 5).

A number of scientists, on the staff of state and federal government agencies, universities, or private research firms, reside in the Willapa region, where they conduct commissioned research. To facilitate and coordinate scientific efforts, the Alliance formed a Willapa Science Advisory Group, which includes scientists and science educators and is staffed by a trained scientist. Not only does such a local advisory group help coordinate local scientific work and science education, but it provides an invaluable linkage to the many outside groups which have a stake in the region's economy and environment, such as government, industry, and universities. Linked to this project is the current work to design and assemble a computerized database, that is, a technical library, on the numerous research projects which have been done in the past on the Willapa ecosystem. The recent hiring of a local botanist as staff director of scientific activities signifies the importance the organization places on good science as a foundation for environmental sustainability planning.

In 1995, to begin to address the social and economic characteristics of the community and to link these to the region's ecosystem, the Alliance published an indicators report, entitled *Willapa Indicators for a Sustainable Community* (WISC). As an initial indicators document, WISC provided an overview of the region's social, economic, and environmental statistics. This document was used as a tool and foundation for the organization to convene a conference in March 1996, entitled Willapa Indicators Leadership Summit. Former Washington State Governor (now Ambassador) Booth Gardner delivered the keynote address; his presence signaled the attention The Willapa Alliance is beginning to receive in the Pacific Northwest as a leader in the development of models for the process of sustainable development. At the summit, over 90 diverse Pacific County residents spent two days in facilitated group discussion to share their vision, hopes, and priorities for their community. The social indicators (WISC) report served as a reference tool for information and focus in their discussion. In support of the Alliance's commitment to help lead the community toward a plan for development which integrates the economic, the cultural, and the environmental, the two-day conference considered and made recommendations for community action in nine areas: culture, health, water resources stewardship, economic opportunity and equity, economic productivity and diversity, natural and built infrastructure, education, land resources stewardship, and civic involvement. The favorable response to the WISC report and to the follow-up community conference has encouraged the Alliance's determination to continue to provide research and information and to convene community members to consider the significance of knowledge about their local physical, economic, and cultural environment.

As a companion document to the WISC report, a directory of organizations and services in Pacific County was also developed in 1995. This document will assist organizations in their individual planning processes and will also help citizens to see the range of places where they can become involved in responding to community issues and needs (1995 Annual Report, p. 4).

Education and Public Involvement

A key challenge facing The Willapa Alliance is to expand the involvement of the general public in its mission and activities. The organization began, essentially, as a leadership organization, that is, a gathering of influential and knowledgeable persons from key industries and other stakeholders in the local region. This gathering evolved into a steering committee and then a board of directors of a formally incorporated organization. The majority of these leaders have a long history and deep stake in the well-being of their local community. Thus, they gradually began to understand and focus on the important goal of increased public involvement in their mission — both for community sanction and for assistance with achieving their program goals. In 1995, a staff person with the job title of communications and education coordinator was hired to coordinate this effort to expand grass-roots (i.e., membership and volunteer) involvement with the Alliance.

To date, education and public involvement activities have been coordinated with the Natural Resource Management and Science and Information programs. For example, The Willapa Science Advisory Group developed several scientific articles which were published in the local newspaper (*The Chinook Observer*). Based on the "Adopt a Stream" concept, local volunteers have been trained to survey streamside conditions and other salmon habitat factors; this information, in turn, was utilized in the GIS and the fisheries recovery plan. Through training and several such field projects, people from a variety of groups (landowners, fishers, timber, government) have been brought together to work cooperatively on their shared stake in the health of the local natural environment.

Conservation-Based Economic Development

Sustainable development is a relatively recent and unfamiliar phrase to some people. To appeal to values and concepts long familiar to most Americans, the Alliance has chosen to emphasize the alternative term "conservation-based development." Like sustainability, the term connotes economic utilization of resources for present needs, with a firm understanding of the necessity of stewardship and conservation of the productive capacity of natural resources for future generations. In spite of its rich heritage of natural resources, Pacific County has a high unemployment rate and a relatively stagnant or restricted situation for economic development. Developing viable businesses, whether large or small, is a major challenge. Both geographic and social conditions contribute to the difficulties. Thus, the Alliance holds conservation-based economic development as a major programmatic goal.

Toward this goal, the Alliance sponsored a local economic development task force. This group in turn helped bring about the establishment of the independent ShoreTrust Trading Group (STTG), an accomplishment with potentially major significance for the local community and region. As is well known in economic development circles, financial resources and technical assistance are essential for entrepreneurs. STTG is modeled on the South Shore Bank in Chicago, Illinois, which has long been known for its principled commitment to lending to prospective businesspeople in its own geographic community, many of whom are low or moderate income. Known as "high-risk" loans to regular banks, many such parties have almost insurmountable obstacles to developing a business idea. South Shore Bank, founded in 1971, has rightfully achieved fame as a model of socially responsible development banking. With the leadership of Ecotrust of Oregon and Shorebank Corporation of Chicago, STTG has been set up as an independent business development institution headquartered in Ilwaco (a small town at the mouth of the Columbia River, south of Willapa Bay). STTG works with socially and environmentally responsible businesses, even though these businesses may be high-risk borrowers by conventional banking criteria. It helps these high-risk borrowers plan their business, practice good management, and find markets for their products. Key criteria utilized by STTG in approval of a loan are that the prospective business must "...be environmentally benign, natural-resource based and hopefully use natural resources that are currently underutilized" (Hunt, 1995b, p. 11). Just as South Shore Bank in Chicago has pioneered the successful application of social criteria to lending, STTG may eventually achieve fame as a successful pioneer in green (environmentally sound) lending criteria (Maughan, 1995).

STTG has two sources of money. Foundation grants of about $2 million and a $450,000 loan from the Weyerhaeuser Company are being loaned out directly for new high-risk businesses. Another pool of funds, called EcoDeposits, totaling about $7 million, has been invested by interested foundations, businesses, and individuals. EcoDeposits offer regular commercial banking services to the depositors (i.e., federally insured savings, checking accounts, and money market dividends). These funds will provide the deposit base for ShoreTrust Bank, a new financial institution which began commercial lending in late 1997 to environmentally responsible businesses with stronger credit capacity than the start-up enterprises supported by STTG (E.C. Wolf, personal communication, October 28, 1996).

STTG has focused on supporting three areas of natural resource utilization: (1) red alder (a native tree which was formerly viewed as a weed tree is actually beneficial to the soil, and its wood also makes beautiful furniture); (2) high-demand non-timber forest resources, such as mushrooms, yew, and ferns; and (3) oysters and salmon from local waters. The organization is also working to help local businesses derive expanded local employment and financial return from local products, through a process known as "adding value." For example, making furniture locally from alder adds to the total economic value which the local community obtains from the wood; smoking or canning seafood locally greatly elevates the price of the exported product. Finding new or better ways to utilize natural products and helping to develop better environmental protection technology, such as better ways to treat sewage, have helped the staff at STTG reinforce the strong commitment by The Willapa Alliance to build its sustainability agenda on a sound scientific foundation. STTG hopes to eventually expand its service area from Alaska to California, a coastal area stretching several thousand miles (Hunt, 1995b, pp. 11–13). Another sustainable development project under consideration is a green industrial park in the region, with potential support from the state government's economic development funds. Meanwhile, as a new local institution, the STTG represents an important accomplishment with possibilities for emulation by other communities.

RELATIONSHIP OF THE WILLAPA ALLIANCE TO GOVERNMENT

As noted in the introductory section of this chapter, the various levels of government (local, state, and federal) have jurisdictional authority and financial leverage over a variety of issues and resources with which the Alliance also is concerned. While members of major industries, such as timber and oysters, have a stakeholder position on the organization's board of directors, governmental representatives do not. In the view of founding members, this was a deliberate organization decision (Arthur Dye, personal communication, July 14, 1996).

However, the Alliance must take governmental authority and interest into account in all of its activities. As one board member put it, "We need to...understand that elected officials and agencies can be good allies" (Jerry R. Gutzwiler, personal communication, October 21, 1996). Many of the environmental and economic development issues of concern to the organization are also of immediate policy-making and implementation concern to various levels of government. For example, the state of Washington has recently passed a Growth Management Act, which requires all counties to prepare a local land use plan, and has already penalized one county in the state which failed to comply. The Alliance also applies its three principles for action to its relationship with governmental agencies. Some members of the Alliance feel that locally generated scientific studies of environmental concerns are more likely to be free of the intense political pressures faced by governmental scientists and at the same time more likely to be responsive to local needs and concerns (Dan'l Markham, personal communication, July 15, 1996).[4] Others believe that local scientific studies face even more intense political pressure to generate specific findings.

The Willapa Alliance's commitment to promoting civic dialogue and partnerships, over approaches favoring confrontation and power struggles, has been applied in its partnership with the Environmental Protection Agency (EPA). The EPA has been mandated to work more closely with local communities on ecosystem projects, but traditionally has not known how to do this, and so was seeking a local demonstration community to develop this capability. The Willapa Alliance aided the EPA with its mandate and in turn was able to help leverage some of the EPA's money for researchers from universities in the region who participated in its project.

Alliance board members feel that relationships with the local county commissioners have been the most problematic. Local officials may indeed have the most to fear from any local organized body of citizens, and the two entities have the closest opportunity to observe and critique one another's position and performance.

In the United States, and especially in the Pacific Northwest, environmental issues are of intense, central political importance. In some localities, the government, especially the federal government, has been identified as the enemy and the oppressor. Mediating organizations which work to find common ground and conciliation of differences among local stakeholders play an important role in working with government agencies to speak authoritatively for local common needs and interests, as opposed to special interests. According to one scientist associated with The Willapa Alliance, it has played a constructive role in public policy debates by developing well-thought-out positions (based on scientific evidence) on the issues (Kim Patten, personal communication, July 15, 1996). Moreover, agents of change must accept that many people do not like change, and they must expect that building people's confidence will be part of their work (Dan'l Markham, personal communication, September 29, 1996).

No one voluntary organization, no matter how broad its stakeholder representation, can claim to represent the entire community. However, this model of locally based consensual decision making on environmental issues is developing in various places in the United States, and it does appear to be diffusing some of the intense confrontational atmosphere between representatives of government agencies, as well as between local stakeholders.

RELATIONSHIPS TO LOCAL NATIVE AMERICAN TRIBES

Native American Chinook and Chehalis (and the Kwalhioquas, who assimilated with both groups in the eighteenth century after European contact) are the largest ethnic minority group in the Willapa watershed. Disease took a heavy toll upon all of these groups during their first encounters with whites in the late eighteenth and early nineteenth centuries. The members of the Chinook Indian Tribe, numbering 1,700, are seeking federal recognition of their tribal status and ratification of treaties from the 1850s that were to guarantee their aboriginal rights to fish and gather in the way of their ancestors. These treaties were never ratified in the nineteenth century. Federal recognition gives tribes a land base, more independence from other governance units, and more access to financial resources. The Shoalwater Bay Reservation, located on the northern shore of Willapa Bay, is home to Chehalis and some Chinook Indians. Native Americans are the largest ethnic minority group and have legitimate claims to environmental resources (especially fishing, their traditional way of life). Their concerns for access to political and social resources (especially healthcare) are also factors which must be acknowledged and addressed in the community's local political and social planning and decision making. Native Americans serve on the board and on the advisory committee of The Willapa Alliance (Gary Johnson, Chinook Tribe Vice Chairman, personal communication, February 10, 1997).

EVALUATING THE ACCOMPLISHMENTS AND SIGNIFICANCE OF THE WILLAPA ALLIANCE

The original stimulus to formation of The Willapa Alliance was the interest of two large, non-local environmental organizations (Ecotrust and The Nature Conservancy). Members of the organization's board, and other local community members, were frank in their acknowledgment of the initial disadvantages this engendered. Coupled with the formation of a board composed of representatives from these outside organizations and from local large business interests (such as the Weyerhaeuser Company and the oyster growers), the Alliance had to work hard to overcome the perceptions of some community members that it was an elitist organization with little accountability to the local community. As one interviewee suggested, some local residents feared that the hidden agenda of the organization was to turn their community into a giant park. The board member from The Nature Conservancy acknowledged that perhaps starting smaller and putting more emphasis on local networking and confidence building might be a better strategy in another community (Elliot Marks, personal communication, July 19, 1996). "Starting smaller" would mean to begin with, and present to the public, small tangible projects and achievements, rather than a long-range master strategy plan (which apparently is what engendered the fear, in some circles, of the organization's objectives). This self-critique is in agreement with classic community organization methods, in which local networking and small-scale victories are cited as the beginning stages of organization building (Rubin and Rubin, 1992). The organization has recognized this key strategic weakness and, with the hiring of a staff member to direct public education and membership development, is taking concrete steps to solidify its position as an organization truly representative of a cross-section of the local communities. "Having a balanced Board that represents all stakeholder constituents is fundamental. This produces more credibility for the organization and provides more and better discussions on technical and policy issues and direction" (Jerry R. Gutzwiler, personal communication, October 21, 1996).

The Alliance's more recent work to study economic and social characteristics of the community may also help balance out its reputation as a group genuinely attempting to integrate the environmental, economic, and social health of the community. A central insight voiced by several founding members of the organization is that people tend to place priority on their economic survival, but that they are motivated to take care of their local physical environment when they can clearly see the intrinsic relationship between their economic well-being and the state of the local ecosystem. The Willapa Alliance has been the object of much interest and attention in the Pacific Northwest, as an organization dedicated to demonstrating and developing these intrinsic linkages. Participation in development of STTG is a laudable accomplishment in the economic portion of this equation. However, with unemployment rates over 12%, and a continued outflow of youth to better employment opportunities beyond the local area, the local population is clearly not in sustainable balance with its ecosystem. Obviously, one voluntary organization can hardly be held accountable for this state of affairs, but it does demonstrate the massive dimensions of the sustainability challenge for local communities. The Alliance may need to consider putting as much explicit scientific and strategic effort into local job creation and retention as it has already directed toward salmon recovery and oyster health; however, these natural resource initiatives have provided a strong foundation to a sustainable jobs future. ShoreTrust's objective to assist with research into environmentally sustainable development and marketing of local resources is an initiative which merits replication by financial institutions in other local communities.

Several factors have made The Willapa Alliance a strong and credible model of sustainable approaches to development. These include:

- Beginning with an identifiable geographic place, a true local ecosystem, which gives people that elusive "sense of place" which is crucial

to human identity and the collective desire and ability to care for that place
- Clearly defining and staying with democratic principles for action, that is, the organization's clearly articulated commitments to fostering dialogue and action partnerships among groups which have frequently been perceived as opponents (environmentalists and businesses)
- Choosing board and staff leaders from the local community with a known reputation for working in the public interest[5]
- Having the good fortune to receive sufficient funding from foundations and other grant-makers to commission solid scientific studies and accomplish notable environmental improvement projects, which have helped build the Alliance's reputation as an organizational leader in sustainability initiatives

The Willapa Alliance is an excellent case model of how a voluntary organization can play a constructive role in bringing together disparate groups in a local environment to address the challenges of building a sustainable model of community.

AUTHOR'S NOTE

I wish to thank all the Willapa Bay community members and members of The Willapa Alliance who generously gave their time for interviews in the summer of 1996. In addition, eight people took the additional time to read and comment meticulously on a draft of this chapter.

REFERENCES

Dietrich, W. (1992). *The final forest: The battle for the last great trees of the Pacific Northwest.* New York: Penguin Books.

Golec, M. (1996). The salmon standard. *The Mountaineer, 90*(7), 8–9.

Hollander, C.V. (1995). Groups cultivate stake in Willapa Bay's future. Special series published by *The Daily Astorian* and *The Chinook Observer.* Willapa: Banking on the Bay, November, pp. 8–9.

Hunt, E. (1995a). Predators thrive in the "pristine" waters. Special series published by *The Daily Astorian* and *The Chinook Observer.* Willapa: Banking on the Bay, November, p. 7.

Hunt, E. (1995b). Solutions for bay mix capitalism, conservation: Lenders focus on "green" loans. Special series published by *The Daily Astorian* and *The Chinook Observer.* Willapa: Banking on the Bay, November, pp. 11–13.

Maughan, J. (1995). Beyond the spotted owl. *The Ford Foundation Report, 26*(1), 4–11.

Rubin, H.J. and Rubin, I.S. (1992). *Community organizing and development* (2nd ed.). New York: Macmillan.

Wilkinson, C. (1992). *Crossing the next meridian: Land, water and the future of the West.* Washington, DC: Island Press.

The Willapa Alliance (1995). *1995 Annual Report* (available from The Willapa Alliance, P.O. Box 278, South Bend, WA 98586).

The Willapa Alliance (1996). *1996 Annual Plan* (available from The Willapa Alliance, P.O. Box 278, South Bend, WA 98586).

The Willapa Alliance (no date). *The Willapa Alliance: A three phase plan* (available from The Willapa Alliance, P.O. Box 278, South Bend, WA 98586).

The Willapa Alliance and Ecotrust (1995). *Willapa indicators for a sustainable community.* August (available from The Willapa Alliance, P.O. Box 278, South Bend, WA 98586).

Wolf, E.C. (1993). *A tidewater place: Portrait of the Willapa ecosystem.* Long Beach, CA: The Willapa Alliance in cooperation with The Nature Conservancy and Ecotrust.

ENDNOTES

1. Several participants in this study objected to this opinion about the aesthetics of tree farms.
2. Chemical methods are part of the process used to control spartina grass, and this receives strong objection from some members of Native American groups (Herb Whitish, Chairman, Shoalwater Bay Tribe, personal communication, October 10, 1996). The problem of how to control invasive (non-native) species is an example of the kind of concrete issues many communities must be prepared to face as they attempt to restore a sustainable local ecosystem.
3. Development of the multimedia CD-ROM project was funded by a grant to another partnering organization, Interrain Pacific, while The Willapa Alliance has coordinated distribution and loan of the disk for use by interested parties.
4. This is a decidedly controversial opinion in the context of U.S. environmental politics, where local control is often a code word for loosening of federally imposed environmental protection standards. A local Native American respondent felt strongly that locally commissioned studies are likely to face more intense political pressure (Herb Whitish, personal communication, October 11, 1996). The controversy demonstrates that science cannot claim total political neutrality.
5. The current chair of the board is a local businesswoman, who is also the daughter of a well-respected local representative elected to the state legislature. Her leadership has lent stature and stability to the Alliance's local reputation.

CHAPTER 11

Changing the Culture and Practice of Development: The Southern Oregon Economic Development Coalition

Joan Legg and Frank Fromherz

> Our mission is to bring together grassroots groups for collaborative action to create an approach to economic development that is financially and ecologically sustainable and to influence traditional economic systems to be more inclusive of diversity in the community.
>
> —The Southern Oregon Economic Development Coalition

Comprised of the working poor, single parents, Latino and Latina farm workers, housing advocates, women, ecological activists, and community and church leaders from Jackson and Josephine counties, the Southern Oregon Economic Development Coalition (SOEDC) is giving *development* a democratic character.

"We are combining our leadership resources and efforts to create diverse micro-enterprises and use community organizing strategies to change local and state policies regarding rural economic development," say the coalition partners. Organizations and groups which comprise SOEDC include:

- Rogue Valley Oregon Action is a grass-roots community-organizing agency with strong focus on leadership development.
- Rogue Valley Community Development Corporation is a broadly based affordable housing organization that builds low-income housing.
- Southern Oregon Growers Association, with a solid history of rural/ farm economic development, is operating a successful growers' market in Grants Pass.
- Convenio de Raices Mexicanas (Gathering of Mexican Roots) is made up of farm workers engaged cooperatively in housing and economic

self-help projects as well as issues affecting Latinos in the southern Oregon community.

- Community Emergency Resources and Vital Services is working with the poor to harvest leftover or surplus crops.
- Peace Garden Institute, a non-profit enterprise, is developing rural land to grow and harvest organic seeds and produce and provide education and training for the community in organic farming techniques.
- Southern Oregon Women's Access to Credit is working with low-income families as they establish credit to build their own businesses and is operating a successful entrepreneur training project to teach business skills and assist participants in developing viable business marketing and feasibility studies.
- American Indian Cultural Center is promoting the interests of non-reservation Native Americans through cultural and economic development and now is developing a center which will give constituents a base to pursue native handicrafts and food-growing techniques.
- Rogue Institute for Ecology and Economy is promoting sustainable forest management practices and assisting businesses as they develop secondary forest products, as well as training unemployed timber workers as forestry resource workers.
- SEEDS, an organic gardening organization, operates a catering service and bakery in Ashland and is involved in developing food processing of natural native plants commonly used by Native Americans and the resident Hispanic community.
- Catholic Community Services of Southern Oregon is educating the community about the diverse needs and talents of the region's people and encouraging the engagement of Christian ecumenical and interfaith religious communities in the mission of the SOEDC.

This chapter tells the story of the SOEDC and how its members are changing both the culture and the practice of development in one particular place, Jackson and Josephine counties in southern Oregon. What does development mean in the mission and action of this broad-based coalition? Development, to be sure, can have various meanings — development of the person, community, natural resources, economy, and so on. This chapter explores the significance of development understood in relation to basic questions of human character: *How ought we to live together, how do we think about how we ought to live together, and how do we put this vision and understanding into practice?* The story in southern Oregon involves questions that tug at the very heart of social and moral ecology and the human effort to sustain a sense of place and the native ecology of a region. Several questions arise from this inquiry into the meaning of development. In the present chapter, for example, we address what role religion plays in an effort to transform both political economic structures and cultural attitudes. We also discuss what participants in the SOEDC are learning from their struggles and accomplishments, their setbacks, and their resilience.

First, the reader will want some sense of the signs of the times in this particular area, Jackson and Josephine counties, known as "the Rogue Valley." As of the 1990 U.S. Census count, some 210,000 people lived in the two southernmost counties in Oregon. Josephine County has the lowest per capita income in the state ($11,438), with 20% of its residents living below the federal poverty level. Jackson County, home to Medford and Ashland (home of the Oregon Shakespeare Festival), rates 21st of the 36 Oregon counties for per capita income ($14,046), with 16% of its residents living below the federal poverty level. The racial composition of the counties is 3.5% Hispanic, 1.5% Native American, 0.8% Asian, 0.2% African-American, and 94% white.

Rogue Valley has the highest unemployment rate in Oregon, 7.6%, and a large number of seasonal workers. Essentially rural in composition, southern Oregon's 60,000 square miles are characterized by clear-cuts, heavily wooded areas, and valleys with small farms and towns. The major population centers of Medford (46,941) and Grants Pass (17,488) are more rural than urban in character and outlook, although certain types of development pressures are beginning to alter this character. The bucolic tone of the region taken as a whole has begun to change significantly.

For a long time the local economy was extraction based — primarily agriculture and timber — until the close of the 1970s. Timber production and housing construction were reduced by the recession of the early 1980s, followed by loss of access to federal old-growth timberlands due to protection efforts to maintain this diminishing ecosystem. The rural areas of Oregon lost jobs at a rate of 4% annually during that period, much of it in high-wage industries such as mining and timber. Per capita income fell and salaries dropped dramatically, so that by 1990 the average wage in Multnomah County (i.e., urban Portland) was 33% more than the average in these two southern Oregon counties.

In response to tough economic times, drawing on deeply established political and economic habits, the most common approaches to economic development in the Rogue Valley have been to expand the tourism industry and to try to attract corporations from outside the Rogue Valley. The paradigm shaping this pattern of development encourages a set of questions along the lines of "What can development do for the economy?" or "What will development do to bolster the economy?" or "How can we get outside investors to participate in our local economy?" With much of the region's job base in tourism-related industries, one pattern is part-time or seasonal jobs that pay minimum wage. This traditional economic decision making tends to approach economic development from a top-down model and a utilitarian cost–benefit analysis. Often the pattern assumes the potential utility of a given place and people, without reflecting thoughtfully on an ecological sense of place, and without asking the character questions that could inform a fuller understanding of human development. Perhaps most importantly, this pattern fails to engage the voices and experiences of people closest to the earth.

Development, in this habitual pattern, tends to be both dehumanizing and disembodied. It loses sight of the true purpose of human development and rarely accounts for a respectful sense of place. The intent is clearly to build a prosperous economy, but the tendency is to reinforce a culture of consumption while undermining ecological sustainability. In a recent survey (Gorski, 1996), visitors to the region actually, for the first time, ranked *shopping* ahead of Crater Lake National Park as the big attraction. In the past decade, Medford's population has increased 34% — 15% more than the state as a whole. Giant retailers have been drawn to the region. A new power center — an open shopping center with mega stores, deep investing, and low prices — dubbed "The Crater Lake Center," paved over an orchard.

The economic relations and cultural attitudes supporting this type of development include, among other things, a short-term profit approach to land zoning, local investments in infrastructure to attract the big investors, and a one-dimensional view of the human person defined as a consumer. It is as if the place and the people have been defined, respectively, by buildability and consumption capacity. Yet the SOEDC offers a challenge to the prevailing culture and practice of development.

BEGINNINGS OF SOEDC

The coalition began meeting infrequently in 1991 and then regularly for the two years preceding the formal development of the organizations to explore mutual problems and develop a common action agenda. Local concerns coincided with a core group under the Campaign for Human Development in the Portland Roman Catholic Archdiocese that attempted to build a coalition in western Oregon (from Portland in the northernmost part of the state through the Rogue Valley in the most southern area). That proved to be too broad in scope. The geographical constraints, with participants separated by as much as 300 miles, proved unworkable, and it was agreed that the groups from the Rogue Valley would be the best instigators of an economic development coalition because of a tradition of working together for mutual support.

The SOEDC entered a planning phase in 1995 to determine what structure would afford the flexibility to change with alterations in the community ecological and economic environment but assure that the structure is sufficient to afford responsibility for fund accounting and operational stability. One of the major questions was "How do we gain the support of the establishment (power) interests in the community without sacrificing our focus and commitment to our constituents, the most powerless and disadvantaged among us?" Without such interests, the SOEDC would remain just another low-income organization with little power or influence. The SOEDC stipulated that the board must be constituted of at least 50% low-income participants. The SOEDC has held workshops on economic literacy and conducts ongoing leadership training to ensure that the low-income representatives do not feel overwhelmed by the more establishment interests.

Through the coalition, members have gained greater access to capital and organizational expertise. The SOEDC has also developed enough political clout to facilitate change in county regulations that are unresponsive to the needs of its otherwise unheard members.

The economic development system in the region, as it exists, consists almost entirely of establishment interests which set the priorities for economic development. It has been made up of people with power in the community who have had little or no contact with the poor. In Jackson County, for instance, decision-makers have set up tourism as an economic strategy, and that tourism brings in service jobs with minimum wage and no benefits. Of course, there is another way to approach tourism, a point to which we will return a little later in this chapter.

For the coalition, opportunities for an alternative approach to economic and human development include light manufacturing, high tech, and agriculture, all done in a sustainable manner. Everything done needs to be environmentally sustainable. "If it's agriculture, it's organic; if it's housing, it needs to use alternative building materials that don't pollute and rape the environment," say coalition members.

COMMITMENT TO ENVIRONMENTAL SUSTAINABILITY

The coalition sees the crisis in the timber industry and the ban on harvesting old-growth logs as a challenge and opportunity rather than the end of viable economic activity in the area. One of the tenets of the coalition is the symbiotic relationship between economic and environmental sustainability. The experience in a timber-dependent economy is that when interests outside of local residents are responsible for utilizing natural resources, the emphasis is on profit, with job creation coming as a by-product. The major focus of the SOEDC is job creation, with profit coming as a by-product. This realization has been with coalition members from the beginning. Self-interest for local residents dictates that the wise use of resources goes hand in hand with job creation, thus developing an economy that assures that jobs are here not only for today but for future generations. Preserving the integrity of the forests and farmland is essential to the economic functioning of coalition members. In their own words, members say:

> We believe that our woods are rich in untapped resources and that traditional forestry interests have been unable to see these resources for the trees. We are blessed with an abundance of plant life growing at the varied elevations in our area. Many of these plants are used in herbal remedies and natural shampoos and cosmetics. Our agricultural interests are focused on cultivating organic fruits, vegetables and berries.

The meaning of development in the purpose and action of the people who have formed the SOEDC provides an alternative to traditional models. Coalition members met informally for several years. In October 1995, coalition members came together formally and announced plans to tap community organizing strategies in order to change local and state policies. The rural community had been severely affected by curtailment of access to federal-owned old-growth timberlands to protect this diminishing resource, as well as by the recession of the early 1980s. The prevalent strategy for economic development promoted by local decision-makers, promoting tourism and attempting to attract corporations from outside the valley, held little promise for improving the condition of the most disadvantaged and powerless. The SOEDC has been designed to unite the efforts of the poor and powerless toward achieving self-sufficiency.

CHANGING THE LOCAL POWER STRUCTURE
FOR DECISION MAKING

Ever since the SOEDC moved into the planning phase of its operation in 1995–96, members have given central attention to power and resource analyses. "Who makes key decisions and how can we access that power?" By going beyond single-issue organizing, by addressing the root disparities of resources and power between the "haves" and the "have-nots" in the community, participating organizations have acted on their conviction and determination to be at the table of economic development decision making in their region. And so they met with and began to develop relationships with the regional planning commission, the Southern Oregon Regional Economic Development Inc. (SOREDI), comprised primarily of powerful members reflecting influential perspectives in the region. The SOREDI lacked the participation of the poor, the powerless, and the culturally/ethnically/racially diverse. This has begun to change.

By combining the principles and practices of ecological sustainability with a commitment to community-based economic empowerment, and by drawing upon years of experience in grass-roots community organizing and leadership development work, the SOEDC is helping to transform the SOREDI into a more representative and collaborative body. The key to this transformation (very much a work in progress) of decision making from a *power over* to a *power among* model has been the coalition's effort to ascertain the most strategic and feasible points for participation by low-income groups. Coalition members describe the power-sharing process this way:

> We discovered that our local planning bodies are required to show how ethnic interests have been consulted in the distribution of public (state and federal) monies. In our area, these planning bodies had been able to address this requirement in a rather vague and generalized way without community challenge. Our coalition, by working together, has been able to direct attention to this matter and require that it be taken seriously. In the appointment to boards we decide upon

a candidate and write letters urging appointment. When ten to fifteen letters are received from a broad spectrum of low income interests and the boards know that we will make our recommendation public, they are quick to comply. The agencies who accept our recommendations find it is a relatively painless process to meet the spirit as well as the letter of the law. We are still working to encourage the boards to exceed the minimum requirements, so far with little success.

The task of changing the habits of institutional decision making has gone hand in hand with the coalition's goal to build strong, ecologically sound businesses. To enable the Rogue Valley community to be more inclusive of the diverse needs and talents of its residents, the coalition successfully placed two of its members in key roles on major county planning boards. The coalition now serves as a member of the SOREDI and participates in its agriculture cluster committee. These efforts to make power more participatory have been accompanied by a concentration on approaches that help people to move toward a fuller vision and practice of sustainable development.

GOALS AND PROGRAM PRIORITIES

In their strategic plan, completed by the end of 1996, the coalition members have decided to devote their combined energies to five areas of development: primary and secondary agriculture, primary and secondary forest products, cultural heritage tourism, health services and related products, and the manufacture and use of non-traditional building materials.

Primary and Secondary Agriculture

As a key emphasis for its first operating year (1995), the SOEDC contracted with the Thomas Group (a technical assistance provider) for a sector analysis of the organic and berry markets, with a particular interest in herbs, which can be profitable on small tracts of land (5 to 10 acres) and can be harvested mechanically. Herbs lend themselves to the organic farming practice being adopted by coalition partners. Preliminary market analysis indicates significant growth potential coupled with a supply shortage. These findings are based upon a number of interviews with both consumers and growers. Government sources indicate continued growth in the size of the market. This suggests a good, competitive environment for organic agriculture. The coalition's plans for the agricultural sector include land acquisition and financing as well as the establishment of a marketing cooperative. The SOEDC is exploring the coupling of value-added products with the actual growing of commodities. One tangible example, developed by the coalition members (SEEDS and Peace Garden Institute) working with the Oregon State University Experiment Station, is the construction of a portable food processor which can be shifted periodically from area to area to avoid the duplication inherent in having each operation construct its own processing equipment.

Primary and Secondary Wood Products

Public timber has become less available and public lands are in need of restoration, yet much less money is available than is needed. Small timber communities that have based their economic lives on timber harvesting are looking for
alternate jobs to replace the limited timber source. Under President Clinton's
timber plan of 1994, some of these timber workers have been trained to do
restoration work in Oregon's forests. These restoration strategies create a new
range of products requiring new approaches to product manufacture and marketing. The coalition has researched a Canadian model log sort yard which encourages diversification among secondary wood products manufacturing. In the
Canadian model, the province sets aside a certain amount of harvestable timber
which is then cut and hauled to the log sort yard. In a survey it became apparent
that there is a local market niche for a sort yard that would be available to small
contractors, artisans, and small manufacturers. The Rogue Institute for Ecology
and Economy, one of the coalition members, is teaming up with Rogue Community College to do further research and to apply this model locally. This
model is sustainable because it uses second growth rather than relying on
protected old-growth stands.

Cultural Heritage Tourism

The traditional top-down model of tourism leads to low-paying jobs and a
superficial understanding of the rich legacies that shape the history of a region.
By contrast, one of the coalition members, the American Indian Cultural Center
(AICC), is developing a partnership with a wildlife rehabilitation program to
develop an independent economic base for southern Oregon's Native population. They plan to develop a site at a state park/rest area along the primary north–
south highway for local and regional Native people to sell products and goods.
The AICC envisions the eventual development of an education center which
would host educational programs that include interaction between Native peoples
and wildlife, with emphasis on Native legend and lore. Eventually the development will include a Native long house and will provide programming to local
schoolchildren as well as tourists.

George Fence of the AICC describes its approach to cultural heritage tourism:

> Our region's geographic location places us at the southern gateway to the Pacific
> Northwest Coast cultural area. This gateway location will serve our efforts at
> previewing Northwest native and wildlife attractions and it is expected to create
> a "marketing service income" while stimulating Northwest tourism and business
> contacts through our marketing effort. The American Indian Cultural Center,
> while working with the coalition, has expanded its partnerships with Rogue
> Community College, Southern Oregon State University, the Oregon State Parks
> and Recreation and is presently courting the United States Forest Service and
> United States Fish and Wildlife agencies. These agencies have been seen by

native peoples as adversarial in the past but present challenges require allied configurations to achieve mutual success.

Health Services and Related Products

The fourth area of community-based development, the coalition's commitment to the health services sector, comes directly from the organizing efforts of Rogue Valley Oregon Action (RVOA) in the health services industry. RVOA, going back to the early 1980s, has a strong tradition of leadership training and community organizing in the region. With the recent increase in the senior population in the Rogue Valley, the medical field has become the largest industry in southern Oregon. There is an increased need for healthcare services. However, as with other aspects of the traditional pattern of economic development, the trend is toward low-paid employment with few benefits. Drawing on the leadership development and community organizing skills of RVOA and building on the accumulated community knowledge of the healthcare economy that RVOA has attained, the coalition has begun to explore the feasibility of a health services cooperative that would provide a sense of shared ownership and family wage jobs.

Manufacture and Use of Non-Traditional Building Materials

Many members of the coalition want to create building components from recycled materials and invest in the construction of homes made from straw and adobe materials. Although these sustainable building projects are still primarily at the talking stage, the diversity of the coalition partnership engenders these types of approaches to economic development and ecological sustainability.

EVALUATING SOEDC APPROACHES

In all of these areas of work and feasibility study, the coalition tries to link community organizing, community-based economic development, and ecological sustainability. Power analysis, power sharing, leadership development, economic sector analyses, and business/job creation are examples of the wide range of skills and strategies used. Of course, no coalitional effort can escape challenges and obstacles along the way. One very strong asset lending cohesion to the SOEDC is the fact that most of the partners already have strong historic ties through the community-organizing and leadership training work of RVOA, one of the oldest and most publicly recognized partners in the coalition.

CHURCH-BASED FUNDING SUPPORT

Like RVOA and several other partners, the coalition itself has been funded by the Campaign for Human Development (CHD). The SOEDC has also received a large grant from Wells Fargo Bank and funding from several other sources.

Attention to the CHD investment affords an examination of several closely related questions. What is the significance of community organizing in shaping the mission and practice of development? How is human development understood in the vision of this CHD-funded project? How does the unfolding story of this coalition in southern Oregon present questions that get at the heart of social and moral ecology and the human effort to sustain a sense of place in the natural ecology of a region? What role does religion play in this effort to engage both economic and ecological change in institutional structures and cultural attitudes?

The CHD in western Oregon, as throughout the United States, works to get at the root causes of poverty and powerlessness. The CHD is a program funded and supported by the U.S. Roman Catholic Church to invest in the empowerment of low-income communities to eradicate the roots of both poverty and powerlessness. Grounded in Christian scripture and Catholic social teaching, the CHD vision is shaped by a faith-based conviction that the poor and powerless — in solidarity with the larger community of persons and institutions committed to greater economic justice and the common good — have the right and responsibility to create their own conditions of dignity (United States Catholic Conference, 1996b).

The CHD vision includes the belief that by organizing, people are going to be more effective (than as autonomous individuals), as they realize their dignity in community. The CHD is an investment of human energies and finances in people finding real solutions to problems of poverty and powerlessness. At the heart of the CHD vision are the Catholic social principles of human dignity and solidarity. Catholic parishioners form the backbone of CHD by their annual contributions to the CHD collection. The CHD's goal is to invest in organized groups of people who have effective strategies and leadership training to allow them to help transform the tables of decision making in southern Oregon.

From this vision, the CHD derives basic core goals applied in southern Oregon as across the country. Working closely with the national CHD, the Catholic Archdiocese of Portland in Oregon CHD invests in community-based projects that have the greatest potential to accomplish the following goals:

- Effect institutional change (get at the root causes of poverty)
- Empower low-income people to gain access to community decision-making structures
- Develop leadership skills
- Generate community cooperation among diverse groups
- Develop ecologically sustainable economic development with business efforts that ensure family living wages, meaningful work, and a culture of respect in the workplace

The SOEDC has received both local and national CHD grants, especially for the purpose of completing the strategic plan. The SOEDC is now seeking a CHD grant to implement this plan.

VALUES UNDERGIRDING SUSTAINABLE DEVELOPMENT BY SOEDC

As the SOEDC has developed its long-range strategic plan, its members have become acutely aware of the differences one's understanding of human development can make in the efforts ahead. "How ought we to live together?" As any community contemplates plans for growth and livability, it ought to ponder the true meaning of human development. Yes, people speak of livability and sustainability, but how might they actually learn to live together for the common good and thus share the good life among all who make a region their home? To this end, the coalition's ongoing understanding of the good life has been pivotal.

The good life in common, genuine human development in mutual relationship with the ecology of the bioregion, requires bonds of solidarity between people committed to economic justice and people engaged in working for ecological sustainability. The good sought in common requires that those who are voiceless have an active place at the table where key decisions shape the future. This the members of the SOEDC believe and practice. Common sense suggests that in the long run economic and ecological health go hand in hand. Too often the people most profoundly affected by economic injustice and environmental degradation have been left out of the discussion. From a justice-minded interpretation of the biblical tradition, many coalition members have been asking who are the widow, the orphan, and the stranger in our midst. Participation, a willingness to listen, and mutual respect have been essential values in the SOEDC view of economic relational justice. No one ought to be left out of the process.

Economic activity and ecological awareness depend upon the relational character of the human condition. The SOEDC members believe that each person and family in their region deserve a living wage. Social ecology as well as natural ecology suffer greatly when so many live on so little, while a few live on so much. The SOEDC shares with the CHD the conviction that everyone has a basic right to a decent living. In this view, rights and responsibilities are relational to the core. Extreme individualism is contrary to both social and natural ecology. In this basic vision of human development, the web of life is one. Every neighborhood and open space, every form of transportation and place of respite, the ecosystem itself and the financial system, all involve relationship. From the science of ecology or from the signs of a depleted ozone layer, many have come to see that no island is entirely unto itself. Actually, some of the most poignant ecological insights have been gained by SOEDC members who have known (e.g., as farm workers) the negative impact of chemical and pesticide-based farming. In effect, the SOEDC has drawn upon a theological and ecological vision, and likewise has recognized the hard-won insights of its own members. These insights have developed from experiences of organizing and from the wellsprings of the faith traditions of coalition members. Traditions represented in the SOEDC, in addition to the Roman Catholic participation already cited, include several mainline Christian protestant and evangelical denominations and Native American faiths.

Coalition members know they have much to learn from one another about what it means to be human and what it means to share a deep and abiding sense of place. As we have described throughout this chapter, they are committed to inviting all participants in the region — including the traditional top-down developers — to rethink the meaning and practice of development.

RELIGION AND ECOLOGY

In 1991, the U.S. Catholic bishops spoke of the inextricable nature of life, economy, and ecology in *Renewing the Earth*:

> Above all, we seek to explore the links between concern for the person and for the earth, between natural ecology and social ecology. The web of life is one. Our mistreatment of the natural world diminishes our own dignity and sacredness, not only because we are destroying resources that future generations of humans need, but because we are engaging in actions that contradict what it means to be human. Our tradition calls us to protect the life and dignity of the human person, and it is increasingly clear that this task cannot be separated from the care and defense of all of creation (p. 2).

In a joint statement from the National Council of Churches, the Synagogue Council, and the United States Catholic Conference (1993) (representing some 100 million American Christians and Jews), religious communities have recognized their own participation in patterns that dehumanize and threaten the way of life: "We acknowledge the complicity of our religious institutions in the brokenness of our society and our world" (p. 84). Acknowledging that the common good is frequently diminished and sometimes destroyed by powerful currents in our world today, they insisted that reflections cannot be "simply an agenda for public policy, but also a necessary guide for wise choices in our families, churches, synagogues, businesses, unions and communities" (p. 84).

If people in southern Oregon can come to agree that the web of life is one, they can likewise agree that it is wise to honor variety within the midst of a unity that shapes both the natural and social ecology. Respectful of the many angles of interpretation and wisdom available through various faith perspectives and secular points of view, the SOEDC is tackling this tough question head on. How might all flourish together as one people in this region, learning to shape their character into the true purpose of human development? How residents and investors come to view and structure development and the common good will have implications for the shape of the future in this region.

A good place to begin deepening an understanding of the common good is found in recognizing common mistakes. Members of the SOEDC, drawing from many faith backgrounds, know that the common good is a calling. It embraces the sum total of those conditions of social living whereby people can more fully and readily achieve their purpose as human beings. From experience in both faith and science, today many would argue that for authentic human development to be encouraged, both the economy and ecology must be sustained. The working definition of a healthy economy will not square, then, with those tendencies that would continue to measure its vitality solely in terms of productivity.

In the view of the SOEDC, land and our thoughtful relation to it offer an ethic in marked contrast to an ethic that views land almost entirely as a utility or a speculative investment. In the social tradition shaping the work of the coalition, people strive to read the signs of the times in the light of the God-given gift of human reason. From the perspective of faith and from the viewpoint of sound human reason, it is not wise to abide so many prevalent signs of dehumanization and signs of fragmentation. Many forces in our society and region tend to emphasize extreme individualism or narrow utilitarian notions of the bottom line, and these patterns left to their own devices — left unaccountable — would undermine both the social and the natural ecology of any region. Such forces must be named, challenged, and overcome by a full community-wide commitment to both economic and ecological sustainability. The point of the coalition's energy is accountability and participatory economics. No one, no corporation, is demonized, yet all must be challenged to define success by terms that a culture of commerce alone cannot grasp.

There are sobering signs of the increasing concentration and centralization of capital institutions in the midst of southern Oregon. Electoral politics and local government may at times be under the control of monied interests. There are signs of increased strife and balkanization as disparate groups compete with each other for a shrinking share of resources. The region has not escaped the national trend toward the worship of an unregulated free market. Fundamentally, the SOEDC is seeking the good in common as its members join forces to halt the widening chasm between the narrow interests of the wealthy few and the common interests of the community. The challenge, of course, is to let go of attitudes and structures which reinforce the tendency to look out only for narrow self-interests. The SOEDC's approach will not be limited to lofty ideals and prophetic words.

Too often families and communities feel that power and influence — and basic security — have been stripped from them by forces that lack a human face. Indeed, often the economic development decision-makers are absent and therefore have little if any sense of local place. Concomitant patterns sometimes take over the approach to the natural environment, as forces of careless growth pave over fertile land and orchards and as natural resources are stripped from us by patterns of excessive consumption.

Such signs are disturbing and they are not without manifestation in the southern Oregon region described in this chapter. It would be more disturbing, however, if people remained unwilling to name and analyze these signs.

Members of the SOEDC are working to encourage the people of their region to listen to the appeal of suffering peoples, to hear the unspoken appeal of nature around them, to be attentive to a spirit of development that is good and genuine, not self-centered wealth sought for its own sake. Gradually, more people in the region are being animated and energized by the SOEDC commitment to the common good. Guidance for this ethic of community character development can be found in the faith traditions that light the imaginations of the people in this coalition.

The shape of the future will depend upon the willingness of local and regional communities to insist that people do not remain shut out, isolated, voiceless, and unable to effect changes in public policies and institutions. Specific action by religious communities and various associations in Jackson and Josephine counties will take many forms, but the SOEDC intends to keep the principle of respectful participation well in mind. If people can remember and be made accountable to remember what it is to be on the path of human development, whole regions may gradually understand and practice the deeper purpose of civic participation and respect for one another.

In this story of the SOEDC, we have tried to account for the coalition's perspective on empowerment and participation, a view of the human person and human dignity shared by the CHD. We have likewise noted the vision of sustainability, a commitment to ecological renewal shared by many traditions including the U.S. Catholic bishops in *Renewing the Earth* (1991). Principles articulated in the statement *A Catholic Framework for Economic Life* (United States Catholic Conference, 1996a) and the 1986 pastoral letter *Economic Justice for All* (National Conference of Catholic Bishops, 1986) point to the theological framework that also is shared by the coalition. Though this statement comes from the Catholic tradition, its sentiments are shared by the many faith traditions of the coalition members:

1. The economy exists for the person, not the person for the economy.
2. All economic life should be shaped by moral principles. Economic choices and institutions must be judged by how they protect or undermine the life and dignity of the human person, support the family, and serve the common good.
3. A fundamental moral measure of any economy is how the poor and vulnerable are faring.
4. All people have a right to life and to secure the basic necessities of life (e.g., food, clothing, shelter, education, healthcare, safe environment, economic security).
5. All people have the right to economic initiative, to productive work, to just wages and benefits, to decent working conditions, as well as to organize and join unions or other associations.
6. All people, to the extent they are able, have a corresponding duty to work, a responsibility to provide for the needs of their families, and an obligation to contribute to the broader society.

7. In economic life, free markets have both clear advantages and limits; government has essential responsibilities and limitations; voluntary groups have irreplaceable roles, but cannot substitute for the proper working of the market and the just polices of the state.

8. Society has a moral obligation, including government action where necessary, to assure opportunity, meet basic human needs, and pursue justice in economic life.

9. Workers, owners, managers, stockholders, and consumers are moral agents in economic life. By our choices, initiative, creativity, and investment, we enhance or diminish economic opportunity, community life, and social justice.

10. The global economy has moral dimensions and human consequences. Decisions on investment, trade, aid, and development should protect human life and promote human rights, especially for those most in need wherever they might live on this globe.

The work of the SOEDC consists of putting these principles into practice. It has long been clear that the pursuit of greater economic justice and ecological sustainability is not carried out primarily by the statements of religious bodies, but in the broader marketplace.

In this story of the SOEDC, we see these principles being put into practice by people trying to transform the culture and practices of development in the process. This story of community-based ecological sustainability could find a higher profile in the public statements of religious leaders from many traditions on economic justice. We hope that grass-roots efforts such as those of the SOEDC may encourage religious bodies to strongly integrate their reflections and actions on economic justice with their statements and actions on ecological sustainability.

The SOEDC has, we believe, a practical approach to genuine human development. We close with some thoughts about the challenges facing the coalition. These words express the collective voice of a coalitional effort:

• Our greatest challenge is also our greatest strength — the fact that we are a coalition. Our board is comprised of organization representatives rather than individuals and so we are assured of an extremely wide base of constituents. This gives us a stronger voice in the community than any one organization could achieve. However, it also means that we are dealing with organizations that can be unstable and volatile at times. We have adopted a policy of not interfering in the internal workings of our members. We have a membership from organizations that are extremely varied in their structures, outlooks, stability, and history. While we learn a lot from one another, often one or more organizations are undergoing crises which require their full attention and resources and are not able to contribute much to the ongoing strength of the coalition.

- We work on the premise that we can make a difference in the lives of powerless/low-income persons through economic development. This requires considerable energy and resources and can be so overwhelming that we are in danger of not producing enough family wage jobs to make much of a difference in the community.

- How much are we willing to compromise our basic tenets to mimic the existing system? Often this becomes the basis upon which we are able to get resources ample to do the job we must do. The biggest obstacle to getting sufficient resources is the education and development of our own local leaders. This requires a tremendous expenditure of organizing energy and resources. While RVOA has been enthusiastic in adopting this task, other organizations are less enthusiastic. This is an important consideration for any area looking to combine economic development and organizing.

- Sometimes our member organizations are so needy that the resources we are able to gather are a drop in the bucket. Economic development is an investment of time and resources on a long-term basis, while many of our member organizations are in need of strategies that promise immediate returns.

- Economic development as a community change strategy is so new that we have little in the way of track record. While we find that community organizing has great effect on economic development, we cannot claim that economic development has a great effect on community organizing. This makes it a difficult sell to community organizers. We have been pleasantly surprised at the reception we have received from the more traditional parts of the community — particularly the local college and SOREDI. However, we need to strengthen our member organizations more to assure that the coalition doesn't fail because of our internal difficulties.

The story as yet untold may yield a harvest of character and courage able to overcome persistent patterns of development that would make of southern Oregon a "rogue" valley, untrue to its ultimate purpose.

Each organization alone, without the collective strength of a coalition, would not have had the capacity to shape a comprehensive region-wide economic development strategy that is ecologically sustainable and democratic (i.e., community based). If each had tried to pursue a vision alone, there would have been duplication of effort. Perhaps the most significant benefit of forming a regional coalition is found in the collective power of many people and organizations speaking as one voice to transform the very meaning of development. The coalition strives to renew a culture and practice of relational respect with the Creator, human beings, and the good Earth.

REFERENCES

Gorski, E. (1996). Medford now finds stores fruitful. *The Oregonian,* December 26, pp. C1, C3.

National Conference of Catholic Bishops (1986). *Economic justice for all.* Washington, DC: United States Catholic Conference.

National Council of Churches, Synagogue Council of America, and United States Catholic Conference (1993). The common good: Old idea, new urgency. *Origins, 23*(6), 81–86.

United States Catholic Conference (1996a). *A Catholic framework for economic life.* Washington, DC: Author.

United States Catholic Conference (1996b). *Campaign for human development: Empowerment & hope, 25 years of turning lives around.* Washington, DC: Author.

United States Catholic Conference (1991). *Renewing the Earth: An invitation to reflection and action on the environment in light of Catholic social teaching.* Washington, DC: Author, November.

CHAPTER 12

Toward Sustainable Irrigated Agriculture in Kazakstan

Mark W. Lusk and S.I. Ospanov

After secession from the Soviet Union in 1991, the Republic of Kazakstan was faced with the daunting tasks of feeding 15 million people, restructuring the economy, and confronting an environmental cleanup of enormous dimensions. The government had left the disintegrating Soviet Union reluctantly, preferring the problems of neo-colonial dependency to the challenges of massive reform. President Nazarbayev eventually put the country into a rapid reform process which will alter the political and economic landscape of Kazakstan for decades. A key element of the reform is the restructuring of agriculture.

Since the early period of Stalin's rule, Kazakstan's agriculture had been organized around large, state-owned collective farms. Often as large as 400,000 hectares, these farms resembled agricultural factories. Employing as many as 5,000 workers, the farms were administered by *apparatchik* administrators who responded to the needs and instructions of Moscow. Production goals, crop selection, agricultural inputs, salaries, and commodity prices were determined by central ministries thousands of miles distant. Not only were all machines, infrastructure, and inputs paid for by the state, virtually all of the production was absorbed by the government. Farmers held highly specified roles such as tractor operator, irrigation technician, or dairy worker. While permitted to operate personal subsistence gardens, farmers otherwise had no say in farm management or cultivation decisions. Essentially, these farms were impossible to sustain over the long haul. They required enormous subsidies, were massively inefficient, and left an environmental legacy of poor soil and water management that may never be fully repaired.

Kazakstan is located in a physical environment in which agriculture is challenging under any organizational form. The country is positioned in the heart of central Asia, bordered on the east by China, on the north by Russia, and to the south by Turkmenistan, Uzbekistan, and Kyrgyzstan. At 2.7 million square kilometers, Kazakstan is the world's seventh largest country. The climate is typical for central Asia, with hot dry summers and cold dry winters. Although precipitation ranges from 150 to 450 millimeters, most of the country is semi-arid. Due to very high mountain ranges to the east and south which capture snow, southern and eastern Kazakstan utilizes mountain drainages for irrigated agriculture. The central and northern regions rely on rain-fed grains and range animals.

Kazakstan is a middle-income developing nation with a per capita gross national product of $1,160. The productive base is declining in most sectors, with the result that per capita gross national product is also dropping (World Bank, 1996). As is true for all of the former Soviet republics, poverty is becoming more widespread and income inequality is increasing.

The environmental legacy of the Soviet period is of such magnitude that it has attracted international attention and investment. Perhaps most widely known is the crisis of the Aral Sea. This sea was once the fourth largest inland body of water, but has declined by over 40% in its surface area and dropped over 14 meters in level over the past three decades due to upstream environmental mismanagement (World Bank, 1993). Basically, the rivers feeding the Aral Sea have been oversubscribed for rice and cotton irrigation and have been contaminated by pesticides, herbicides, and fertilizer. The fishery in the Aral Sea has disappeared and the region around the sea is affected by salt dust storms. The same phenomena are now beginning to affect the Caspian Sea. International donors are investing in irrigation reform to increase the freshwater inflows and reduce contaminants to the Aral and Caspian seas.

Elsewhere in northeastern Kazakstan are nuclear dumps, abandoned nuclear test ranges, and missile sites. Epidemiological data reveal high cancer rates and genetic abnormalities in the test region (World Bank, 1993). The nation also faces severe urban air pollution, significant water pollution, an accumulation of hazardous solid wastes, declining forests, and diminished wildlife resources.

Although much of the environmental mismanagement of Kazakstan's resources originates in the industrial sector, agriculture has been a major contributor due to water-consumptive crop choices (rice and cotton), poor drainage, and the overapplication of agricultural inputs. Agriculture accounts for about one-quarter of gross domestic product and 23% of total employment (World Bank, 1995). Major products are grains, potatoes, fodder, meat, and wool.

With the challenge of competing in a global agricultural economy and providing for its own people, Kazakstan had privatized all of the state farms by mid-1996, at least on paper. But, bereft of capital resources, the central government has been unable to provide technical assistance to farmers as they make the transition to a market economy. Moreover, poor farming and maintenance practices had seriously depleted the agricultural environment, thereby limiting its productive potential. Faced with the prospect of continuing decline in agricultural production, the government of Kazakstan has aggressively sought technical assistance to turn the situation around and make the transition to a free market system of agriculture. Also, recognizing the perilous condition of its environment, the government of Kazakstan has been striving to reform food production away from a consumptive use of the environment to a sustainable one.

This chapter summarizes agricultural reform in Kazakstan from 1991 to the present. Briefly discussed are the challenges of economic reform in a post-Soviet society, the nation's agricultural privatization program, and the path to agricultural reform with emphasis on sustainable irrigation. The present study was undertaken as part of a national initiative to reform the management of agricultural irrigation and drainage in Kazakstan as the country seeks to move from command agriculture to sustainable agriculture in a private economy. Of key importance to the research is the goal of developing sustainable agricultural water management models appropriate for use in the nations of the former Soviet Union and other developing contexts.

BACKGROUND

Agriculture in Kazakstan is moving from centralized state control and production to a more diversified pattern. In 1992, there were 2,100 state collective farms with a mean of 80,000 hectares each in area on average, only 16% of which was cultivated. As a result of the agricultural reform process of the past few years, all of these farms have been formally privatized, although this process has been uneven in the degree to which the former state farms genuinely reflect the characteristics of private, market economy farming systems as these are normally understood.

New forms of farm organization have emerged. Among them are *collective farms* in which the property rights have been subdivided. Most commonly this subdivision involves stock shares which are distributed to farm workers on the basis of number of adults in the family, length of service on the farm, and farm job position. In other cases, the subdivision has been informal, and the farms continue as collective enterprises with essentially the same organizational framework as before "privatization." These farms also largely still function with the same distribution of labor as before; farm workers continue to cultivate under brigades, farm experts continue to select crops and rotations, and farm managers and engineers continue to control most aspects of the farm. In such schemes, it is difficult to identify the farmer per se as farm labor is still very much associated with individual worker roles such as mechanic, tractor driver, irrigation worker, and so on.

Some shareholders on collective farms have broken away and converted their shares into discrete farm units which they privately and independently administer as a *private farm* or *family farm*. Under law, the land itself remains the property of the state; however, the farmers acquire inheritable use rights over the land. This is equivalent to permanent lease rights which allow them to cultivate as they wish, rent the property, and sell or bequeath the lease. Some private farms are precarious because of the difficulty of purchasing farm machinery and inputs as well as access to markets and credit. Some private farmers have spontaneously organized into informal farmer associations. This process continues to be a fluid one and redefinition of farms is ongoing.

Access to markets constrains the degree to which farms can fully privatize. While commodity producers near urban markets are in a much better position, other farms relying on crop or dairy production are working against market factors which preclude their profitability. In Chengeldy, for example, the former state farm has cut its cattle production by 80% as the per kilo price of meat fell below the production cost. In addition, with the shortage of cash, the farm has been forced to pay some of its bills by bartering off crops and cattle. Farms report a marked liquidity problem and have expended many of their non-essential assets as barter for inputs (Euroconsult, 1995). Limited funds for fuel for machinery reduce the capability of producing maximum yields, and fertilizer is too expensive. Maintenance of irrigation pump systems is uneven, and pump failures are not uncommon. The high price of credit is particularly acute; farmers have spoken of interest rates ranging from 100% to over 300% annually. Some sentiment in favor of the unrealistic goal of returning to state farms has been observed. The lack of short-term profitability in some systems will constrain enthusiasm for the privatization process from some of its potential beneficiaries. While many of these problems can be overcome with effective training for farm business management, realistically, for privatization to be successful and sustainable, it must be accompanied by free market pricing mechanisms, increased investment, and training.

PRIVATIZATION

Although large-scale privatization is comparatively new to Kazakstan, it has been rapidly proceeding in other formerly state-command economies. These economies, in which as much as 90% of economic activity was accounted for by state-owned enterprises, have been proceeding rapidly toward large-scale privatization (Nellis, 1991; World Bank, 1996). Important parts of overall reform include monetary deregulation, fiscal reform, the lifting of price controls, and the elimination of both internal and international trade barriers. Agriculture, which in most countries has been subsidized and protected by tariffs and trade barriers, has been increasingly subjected to free market forces worldwide, although this is less prevalent in the former socialist states. In all, more than 80 countries have privatized state-owned enterprises.

The challenge of privatization in Kazakstan, whether of industry, agriculture, or natural resource management, is to overcome weak legal frameworks, limited financial and auditing structures, poor access to reasonable credit, and limited social capital. If we define *social capital* as the civic culture which supports organizational development with trust, open decision making, and an emphasis on the communication process, then it is apparent that the former Soviet economies have much to do in this domain as well (Fukuyama, 1995). Additionally, weak legal frameworks for the market economy and a shortage of liquid capital constrain the rate and success of privatization. However, because the Kazakstan agricultural system is rich in its ability to mobilize labor and exact organizational obedience, there are many strengths with which to work.

Agricultural Privatization in Kazakstan

During the past five years, all of Kazakstan's 2,100 state farms have been "privatized" (World Bank, 1995). The formal part of this process proceeded in stages, including property appraisal; determination of shares to be distributed to farm workers; intra-farm consultation to establish the basis for a new farm structure; ratification of the organizational structure by all members of the farm in a general assembly; and the election of a manager and, in some cases, a board of directors (World Bank, 1995).

Agricultural reform in Kazakstan has proceeded under a series of presidential decrees enacted from 1991 to 1994. Although to date, *land still remains in sole possession of the state,* it can be transferred to individuals or economic units in the form of a lease of land tenure which is legally permanent and inheritable (BRIF, 1995a). Interpretation of these decrees varies by farm and *oblast* (state). Some farms have broken off from former state farms, acquiring their land free of charge based on a farm worker's acquired share allocation. These units function as private farms, but many do not have a title or certificate to the land. In some cases, farm workers who sought to secede from the collective farm have not been permitted to do so by the farm managers or directors. It is evident that practice varies widely.

The privatization of state farms has resulted in the emergence of several types of farm organizations including joint stock companies, farmer cooperatives, comrade associations, and private enterprises (World Bank, 1995). The most common collective form is the joint stock company in which farm workers are granted shares based on such factors as length of service, occupation on the state farm, and others. This form retains many of the elements of a state farm, including strong central management, production quotas, brigade labor, collective ownership of machinery and land, and the collective sharing of profits and debts. In these farms, individual families are permitted to operate small gardens and to possess livestock for personal use. These personal resources are a major factor in the subsistence of the farm family.

Cooperatives function much like joint stock companies except that the managers of the farm are elected by members. Although several organizational types fall into this category, the former administrative structure is usually preserved along with the farm *nomenklatura* of experts, including engineers and agronomists. Members of cooperative or collective enterprises may leave the farm and establish small private farms with land holdings appropriate to their share of stock; it has been observed that lands they acquire are generally inferior, often located at the periphery of the farm, and have limited access to irrigation or technical assistance.

Excepting small private farms, most farms still look much like state farms in that workers exercise little or no control over farm decision making in cropping, production targets, inputs, and irrigation. Since the situation is highly dynamic, we can anticipate new forms of organization as well as shifts in the above categories. The family farm, in particular, is a rapidly growing type of farm structure.

Privatized state farms in Kazakstan are moving toward employee ownership. The joint stock company, in which shares are distributed only to workers and managers, is a limited voucher-type program or closed offering joint stock company. Presumably, these vouchers, or shares, can be converted into individual land use rights or retained as shares to entitle the owner to a proportional allotment of decision making, machinery use, profits, and losses. This first step in mass privatization, usually referred to as "corporatization," entails the process by which a state enterprise assumes its own legal identity, becomes an open (or limited) joint stock company, and selects a board of directors that "has the right (in theory) to appoint and dismiss management, and operates under the same commercial law as the private sector" (Karlova, 1995, p. 245). In contrast to employee buyouts in Western economies, the corporatization of agriculture in Kazakstan has been implemented by senior managers under central state mandate, thus accounting for some of the lag and limited enthusiasm for the process by workers and shareholders. Theoretical economists had predicted that employee ownership in the former Soviet Union would result in the diversion of profits to worker wages and benefits, limited investment in capital improvements, and worker domination of the board of directors. Research in Russia indicates that this prediction has not been borne out and that employee benefits have actually declined as firms are subjected to competitive pressures (Blasi, 1995). In Kazak agriculture, management entrenchment in farm enterprises is such that employee control is still distant, and the anticipated benefits of employee ownership (profit sharing, decentralized decision making, and higher productivity) are not yet a reality.

Irrigation Privatization in Kazakstan

One review of irrigation reform (BRIF, 1995a) concluded that:

- Water supply was better before privatization. There were problems of maintenance before the reorganization, but they are worse now because they must be repaired with farm money rather than state money.
- The majority of farms have not paid off their debts, and this is accelerating irrigation system decline.

- Equipment use has been problematic due to its uncertain ownership; some of it has been stolen, and in some cases, sowing has stopped.
- Shortage of finance is a particular problem.
- Concern was expressed by managers and farm workers over the rising price of water.
- Although most farms have been privatized, most cannot be considered to be working completely on a new basis, and for many workers the rights to shares and dividends are still theoretical.

Few, if any, collective or stock company farms have chosen to privatize irrigation and drainage systems, reportedly because of lack of money. Also, because land and water are state property, it is assumed that irrigation and drainage are state responsibilities (BRIF, 1995b). Because it cannot be expected that farmers will make capital improvements and investments in farms in an uncertain legal environment in which there is an absence of clear tenancy and property rights (Lusk and Parlin, 1991), it is important to clarify ownership and the authority and responsibility for irrigation and drainage. Yet at the same time, it is clear that operation and maintenance (O&M) of irrigation systems involve capital costs as well as labor, and while workers may be willing to share in labor through a *corvee'* system, they must establish revolving accounts for the capital costs of O&M.

In field visits, we observed that where boards of directors are in place, farm worker representation is uneven or absent. In addition, managers generally market commodities to state agencies and complain of lack of direct access to markets due to transportation and fuel costs. In a focus group conducted on a joint stock company, one farmer observed:

> The structure here is the same as we had under socialism. This system was good at planning, but not at management. Now, in an unstable economy, it is difficult to manage such a large farm. We have no assistance in on-farm water management techniques. Prosperity will not follow from increases in water and inputs to the farm. Credit should go directly to the farmers according to the shares that were distributed to the shareholders. Only in this way would an international loan be efficient.

When confronted with a new reality that has cost farm productivity and profitability over the short term, initial farm member pessimism was to be expected. But our research did reveal growing enthusiasm for the privatization process over the two-year period 1995–96. A long-standing way of life is being turned upside down and few supports have been in place to ease the transition. If, in contrast to maintaining the status quo, farm members are involved in the design and layout of irrigation systems that serve the varied constituencies and units downstream from the head gate, and if the costs they are expected to bear are supported by a strong and free market for their products, we can expect their views to change rapidly. A vehicle for such change is the private irrigation company.

WATER SHARES

Clear ownership, tenure, and rights of use are central to efficiency in market economies; in their absence, individuals are unwilling to make capital improvements and take risks. This is no less so in an irrigated agricultural environment. Shares are the means by which ownership and tenure of water use rights are established. Broadly defined, every irrigation system that serves multiple users is based on private water shares. They may be informal and based on consensus, as seen in some village schemes, or they may be highly formal and based on legal shares or certificates, as is common in mature schemes. A share system reflects the distribution of assets in a given social system. It can be egalitarian or centralized and oligopolistic. Share systems determine irrigator rights by defining access, volume, timing, cost, and rules of water distribution. Shares are the social engineering of irrigation and can be designed to fit local circumstances and cultural traditions. If irrigation water is running without a formal or informal share system in place, the results can range from irrigation anarchy to total domination by a powerful subset of users.

Share systems can be organized in many different ways. These include primacy rights for early users, primacy rights for subsistence users, rotations based simply on land area, rotations based on number of shares bought or traded in the market, shares held for fisheries, correlative rights based on the safe yields of limited aquifers, hierarchical shares based on total sustainable system volume, and mixed systems containing one or more share rules in combination. Shares can be organized by volume, timing, location, seniority, or land area. If the system is to be efficient and receive social support for operation, maintenance, and fee payment, the ideal manner in which to organize the shares is to allocate water based on a proportionate volume equivalent to the legally appropriated right of use. However, without irrigation institutional rules which are widely supported, understood, equitable, and enforced, even the best designed system will quickly fall into disrepair and irrigation deviance will become widespread (Freeman, 1991).

Irrigation water, as a private good, is not free or *gratis*. There are real costs associated with its collection, storage, diversion, and distribution that must be accounted for in the overall management of the system. Because food security serves national and regional interests, it is common for the initial costs to be borne in large measure by the state and region and to capture other rents from domestic and other uses when feasible. It is desirable for the ongoing O&M costs to be shared among the actual users, so that these costs can then be passed on to consumers of agricultural products. This promotes environmental management by incorporating conservation incentives. As with all other aspects of irrigation organization, these costs and the methods of deriving them should be transparent in order to garner support and engender trust in the system.

In a former state farm scheme, the ease of distributing water by volume according to stock shares is apparent. One need only compute total annual sustainable system volume and divide by total farm shares to determine the appropriated volumetric share. This becomes the bottom line against which O&M fees can be levied.

IRRIGATION DEMOCRATIZATION

While much has been written about changing farmer behavior in irrigation (cf., Korten, 1982), some attention has also been directed toward bureaucratic reform in water management (cf., Borlaug, 1987; Bromley, 1982, 1987). A central theme has been the democratization of irrigation. "Democratization is the process of building (representative) political accountability into organization design" (Lusk and Parlin, 1991, p. 27). The democratization of water management provides for accountability to users and funders and scales decision making back to the primary constituents. In centralized bureaucracy, the risks of non-accountable decision making are great and can lead to corruptibility and abuse of authority. In part, this process is accompanied by scaling down decision making (decentralization) because larger groups tend not to induce as great a sense of accountability as smaller, more transparent groups which involve face-to-face relationships. Smaller decision-making groups induce a sense of reciprocity, mutual accountability, and the emergence of trust — especially when group members expect to have long-term relationships that require collaboration (Fukuyama, 1995).

Irrigation companies and water user associations tend to support democratization when the bylaws subject management to water user oversight. These organizational forms can break out of top-heavy hierarchical modes and are more responsive to market agriculture, which requires flexibility, good information, and cost efficiency. It is naive to appeal to the altruistic motives of leaders or to "bureaucratically reorient" them to working in the public interest (cf., Korten, 1980; Bagadion, 1985) and far more important to incorporate incentives for success and accountability for failure into the reward structures that guide bureaucratic behavior. The water user association is an effective mechanism for building such responsiveness into leadership behavior. The design of representative bureaucracy can be built around the election of farm directors and irrigation employees. When farmers are seen as the stockholders of a company, their water use rights and entitlements are adjudicated and transparent, the rights and responsibilities of water users and their elected officials are codified, water use is optimized through fees set by pricing mechanisms, and decision making is decentralized and democratic, success follows.

RESEARCH ON IRRIGATED AGRICULTURE REFORM IN KAZAKSTAN

Ten large newly privatized farms[1] and state farms scheduled for privatization in seven *oblasts* were included in our national study. Twenty-eight groups were studied: ten semi-structured interviews with the managers of restructured state farms, ten focus groups with farm workers[2] and specialists, and eight focus groups with independent farmers. In all, 280 people participated in the discussions. Field work included farm reconnaissance and inspection.

We sought to understand how irrigation is presently administered; determine if farm workers, owners, and operators have irrigation access proportional to their share; discover what water use rules and regulations are presently in effect; determine if water user fees are presently charged, and if so, on what basis; explore attitudes toward the private management of irrigation water; determine water users' willingness to pay for O&M; find out who controls cropping and inputs; explore what technical assistance is needed in irrigation organization and on-farm water management; and explore support for alternative forms of water user organization.

GENERAL FINDINGS

The transition of Kazakstan's agriculture has been fraught with great difficulties. Agro-enterprises reported that yields on all major crops and products are down markedly since 1991. Farm infrastructure was found to be in deep disrepair. Many irrigation control and distribution structures are non-functioning, with the result that total water supply is down significantly. Some machinery and parts have been stolen or sold. The availability and costs of parts as well as limited access to affordable credit are such that repairs are not possible at this time.

Total area under production has also declined not only due to the disrepair of the infrastructure but also because of the deterioration of soil productivity through erosion, salinization, waterlogging, and the lack of funds for needed inputs such as fertilizers.

Farm incomes have decreased on newly privatized farms, and the standard of living has been adversely affected on joint stock company farms and cooperative ventures. Smaller independent farmers are actually doing better than before reform. Small farmers reported that because they could make all of their own decisions about cropping and marketing, they were able to produce surpluses for the market and raise their incomes. Because they do not rely on industrial agriculture techniques and economies of scale, they are able to work more efficiently.

Generally, market prices have not kept pace with the rising costs of inputs, and many former state farms have resorted to paying their bills with barter or not at all. Not only have lands been taken out of production because of the price/cost squeeze, but dairy and breeding herds have been sold off below cost or slaughtered.

Many of the difficulties of the transition period correspond with the delayed effects of the Soviet period, which in the absence of large state agricultural subsidies became manifest. These include highly centralized decision making, deferred maintenance, and controlled market prices. But the elements of reform, although painful at first, are well under way to being realized. Separate, independent farms have been created often against great odds. Independent farmers reported that farm managers[3] and *raion* (county) administrators have not facilitated either the secession of independent, private farms or their success, and yet many farmers continue to elect to pursue an independent path. To fully incorporate them into the potential gains in productivity from agricultural reform will necessitate the inclusion of independent farms in autonomous farmer organizations which control the irrigation system.

SPECIFIC FINDINGS

Water Supply and Control

The supply of water within the large agro-enterprises is still controlled by hydro-technicians and managers. Workers report that their sole influence over distribution, volume, and timing is through brigade leaders. *Independent farmers do not have influence over its distribution, volume, or timing.* On-farm water management is constrained by limited supply to the farm from the Raion Committee on Water Resources. Although farm allocation is determined by law, most farms receive a fraction of their allocated water requirements and report that they obtain between 50 to 90% of planned volume. Managers and specialists[4] report that shortage forces them to take fields out of production or to underirrigate, which reduces yields. Many independent farmers are located either at the tail of the system or outside of it altogether and cultivate rain-fed crops or rent arable land from the farm.

Water is controlled by the farm managers, hydro-technicians, and brigade leaders. Interviewees who work at the field level reported that despite central control, water is not generally delivered on schedule. With rare exception, there are no water-measuring devices and thus time rather than volume is used as the criterion. The deferral of maintenance has resulted in some parts of the farms receiving no water whatsoever.

Water Shares

In theory, workers on newly privatized farms have access to the irrigation system in proportion to their shares in the farm. In practice, this translates into the ability they have as members of the brigade to influence on-farm water management. Practically speaking, shares in joint stock companies have not translated into water shares. The assets (land, equipment, etc.) and debts of the enterprises have been allocated among the members in the abstract rather than in kind. Presently, an independent farmer cannot lay claim to a water share or a tractor for individual use.

Water Fees

The Raion Department of Irrigation Systems determines the prices of water at farm head gates. In the past, the calculations for water fees were presented to the farm administration, but this practice is no longer followed, and pricing is not transparent. Prices to agro-enterprises have increased significantly over the past three years. Individual shareholders or separate farmers do not directly pay for water. Our data indicate that for the past water year, on the ten farms in the sample, a total of 23.3 million tenge (75 t = US$1) was owed and only 3.6 million tenge was paid, reportedly because of the shortage of cash and barter and because the volume delivered was less than the allocation. Evidence of arbitrary pricing was seen, yet managers said that they felt it unwise to challenge water prices or they might face reprisals such as increases in fees or reductions in volume. In seven out of ten farms, farm workers said that they had no idea how water fees were calculated. Independent farmers had no knowledge of water fee calculation.

Independent Farms

Within the study region, about 325 small independent farms have been formed. Only a fraction of the separate farms have been assigned irrigated land (about 10%). Lands received under privatization are inferior, with some exceptions. Lands are often pasture, waterlogged, brushwood, or have never been cultivated.

Private farmers said that they have been virtually ignored by the administrations of the newly privatized farms. As noted, they have been allocated generally poor land that is often outside the irrigation scheme. In some cases, they said that impediments have been placed in the way of their formation, and once established, they have tended to function outside of the sphere of operations for the larger agro-enterprise. They generally receive no share of the equipment and have not been included in the areas cultivated by the larger agro-enterprises. None of the separate farmers in our study area have engaged in any off-farm economic activity. Their knowledge of how to work in a private agricultural economy is limited.

The time to acquire a private plot was reported as lengthy, between three months and three years. Access to state credits is not available and private credit can be at rates in excess of 300% per annum. Interestingly, *private farmers generally reported that their welfare standards were higher than those for workers in restructured state farms.* In all of the study areas, private farmers reported that their economic situation had improved as a result of privatization. They observed that they were able to pursue their own self-interests rather than those of the larger farm and that they were able to produce those crops or animals they deemed most profitable. Despite the difficulties faced, they responded favorably when asked if privatization had been a good thing for Kazakstan's agriculture. When private farmers were asked if privatization had been a positive thing for Kazakstan's agriculture, 75% said yes and 25% said they did not know. No one expressed a desire to return to state agriculture.

The challenges private farmers face are lack of marketing knowledge, limited access to markets, access to credit, irregular or absent water supply, and no access to extension training on irrigation technique, agronomy, marketing, computing, and accounting. They want to purchase small farm machinery and participate in the organization of irrigation water.

Conflict Management

Interesting variations in conflict resolution were reported by farm managers and independent farmers. Managers tended to respond that there has been little or no conflict on the farm or that when there was it was handled "in the working order." Private farmers said that conflict had been handled among themselves. When independent farmers had a conflict with farm management, they said they were ignored and thus the conflict "disappeared." It should be noted that one of the primary functions of a water user association is the formal resolution of conflict among and between members.

Choice of Crops and Inputs

On restructured state farms, managers and experts generally select crops to be grown, areas to be cultivated, and inputs to be applied. In some cases, farm workers are involved to a limited degree in these decisions. On private family farms, individual farmers make these choices, but they generally cannot afford inputs and select those crops with which they have familiarity.

Stakeholder Attitudes

Potential beneficiaries were generally positive about the prospect of new models of irrigation management. While they reported uncertainty as to what forms such structures might take, they anticipate that a change in the status quo is desirable. This opinion was predominant among independent farmers and to some extent among workers, but, for obvious reasons, farm managers were not as positive about the benefit of farmer participation in irrigation management through water user associations. In one manager's words, "it is undesirable, the old rules should be kept." Individual farmers were more positive and expressed an interest in joining a water user association if it did not increase water costs and did not result in monopoly or corruption.

During a USAID/Harvard Institute for International Development–sponsored National Seminar on Water User Associations in Almaty in 1996, in which the authors were facilitators, overwhelming support for a national effort to form water user associations throughout Kazakstan was generated among officials of the Ministry of Agriculture, representatives of all of the nation's *oblast* governments, and farmer delegates. The group endorsed a resolution which supports:

- Legally recognizing the water user association as a basis for organizing irrigation at the local level
- Nationally funding the rehabilitation of deteriorated irrigation infrastructure
- Normalizing water fee structures
- Carrying out pilot projects in water user association formation
- Decentralizing and democratizing the nation's irrigation management structures

In our research, a majority of farm worker focus groups (60%) reached consensus on supporting the idea of paying irrigation fees for maintenance and water delivery if it improved the timing and volume of water. Participants felt that it was important that pricing not be arbitrary and that it be tied to crop, yield, volume, or some other reasonable criterion.

We found that workers and farmers want greater participation in the management of water. They want greater predictability in the timing and volume of water allocation and greater assurance of predictable patterns of delivery. This can be accomplished through some variant of the water user association being formed in each of the study areas.

Although wanting greater input into irrigation management, farmers did not know what water user associations are or how they are formed, managed, funded, and held accountable. Focus group discussions of organizational alternatives were thus quite preliminary and covered the basic features of farmer water user associations. The focus groups also revealed that stakeholders have positive attitudes toward the privatization of irrigation systems. But it was also evident that they did not fully realize that they themselves would become the owners of the system, not a new board or bureaucrat.

It is evident from the research that the ground is ready for new institutional forms of irrigated agriculture in Kazakstan. The organizational alternatives for management under different types of irrigation systems differ minimally. In each case, a water user association is formed for a discrete hydrological subproject area and fit to the technology. Each water user association takes responsibility for the system at the headgate or pump and manages the water at and below the intake point. This organizational form can assure representation and foster conservation and sound water management.

SUSTAINABLE IRRIGATION ORGANIZATION: THE WATER USER ASSOCIATION

Independent private farmers have shown early success as they have made the transition from state industrial agriculture to private small-scale farming. They have improved their living standards, increased their farms' productivity, and produced surpluses for the market. This has been despite enormous obstacles. The next step in securing their future is to begin to form farmer associations for cooperatively marketing their goods and managing their water resources. These farmers have moved from complete state domination to disaggregated agriculture. While this has been initially beneficial, they have not achieved the economies of scale that could come from cooperative purchase of inputs, management of machinery, marketing, and organization of irrigation water. The next challenge is to fill the institutional vacuum created by the ongoing collapse of command agriculture.

The private water user association is a medium-size organization that can serve many of the functions needed to fill the void left by the end of the state farm. The water user association is a corporate form of organization which entails an elected board of directors, a manager, irrigation employees, and a definition of water users as stockholders. It is governed with bylaws and is by definition a representative organization in which the bylaws and management of the organization are controlled indirectly by the water users themselves. It stands in contrast to the bureaucratic irrigation organization which controls water centrally with little direct accountability to the local farmers. The water user association can serve not only to manage water, but can also provide the organizational platform for cooperative commodity marketing and purchasing of inputs.

This organizational form has a precedent in the joint stock company in Kazakstan and is the easiest way to organize water because of this precedent and the widespread understanding of the corporate/collective farm. It is also applicable to Kazakstan because the water user association views water users as discrete elements, thereby allowing for the inclusion into the membership different forms of farm-level organizations from the large farm cooperative to the private family farmer. Each type of user can be assured of representation on the basis of stock shares and can claim water in proportion to those shares on a volumetric basis. During droughts or scarce water years, a portion of the total reduction of farm volume can be shared by members in proportion to their cultivated land area.

Antecedent to the water user association is the requirement that area bureaucrats be prepared to recognize the water user association and grant it the authority to control inter-farm water within the jurisdiction of the water user association (Freeman and Lowdermilk, 1981). A second condition is that the farmers must be allowed to cover the cost of the centrally delivered water based on fees assessed in proportion to their shares, area, and/or crop.

When water user associations are built into the management of irrigated agriculture, farmer participation in what had previously been a state agricultural system becomes central. With greater constituent involvement in the control of the scarce natural resource, the sustainability of agricultural reform can be enhanced.

SUMMARY

Agricultural reform in Kazakstan has been proceeding rapidly, but unevenly. As the most productive lands in the country are under irrigation, a key component of agricultural reform has been the restructuring of water management. This chapter has summarized agricultural reform in Kazakstan, noted the early successes of the small-scale farmer, and identified those elements of irrigation reform which will be necessary for this transition to be completed.

Among other things, it was found that farmers and other stakeholders in agriculture generally support agricultural privatization. They also support the privatization of irrigation management and are willing to pay more for water if it will improve the volume and scheduling of supply. Importantly, it was also found that the private, independent farmers who have broken away from newly privatized state farms enjoy welfare standards higher than those of members of joint stock companies and collective agricultural enterprises, and for them this has been a remarkable accomplishment. These findings bode well for the prospect of completing a genuine shift toward market agriculture. Perhaps the greatest challenge in sustainable agricultural reform will not be in mobilizing local support for reform and privatization, but will be in breaking down the macroeconomic barriers to private agriculture such as pricing policy, access to external markets, and bureaucratic resistance.

AUTHORS' NOTE

The authors gratefully acknowledge the support of the Harza Engineering/World Bank Kazakstan Irrigation and Drainage Improvement Project and the Harvard Institute for International Development Environmental Economics and Policy Project.

REFERENCES

Bagadion, B.U. (1985). *Farmer participation in irrigation management in the Philippines.* West Hartford, CT: Kumarian.

Bell, S.W. (1995). *Sharing the wealth: Privatization through broad-based ownership strategies.* World Bank Discussion Papers, No. 285. Washington, DC: The World Bank.

Blasi, J. (1995). Ownership, governance, and restructuring. In I.W. Lieberman and J. Nellis (Eds.), *Russia: Creating private enterprises and efficient markets* (pp. 125–140). Washington, DC: The World Bank.

Borlaug, N.E. (1987). Making institutions work — A scientist's viewpoint. In W.R. Jordan (Ed.), *Water and water policy in world food supplies* (pp. 387–396). College Station, TX: Texas A&M University Press.

Boycko, M. and Shleifer, A. (1993). The voucher program for Russia. In A. Aslund and R. Layard (Eds.), *Changing the economic system in Russia* (pp. 236–248). New York: St. Martin's.

Boycko, M. and Shleifer, A. (1995). Next steps in privatization: Six major challenges. In I.W. Lieberman and J. Nellis (Eds.), *Russia: Creating private enterprises and efficient markets* (pp. 75–86). Washington, DC: The World Bank.

BRIF (1995a). *Report on the focus group study: Social assessment of stakeholders in irrigation rehabilitation and farm privatization in Kazakstan.* Almaty, Kazakstan: BRIF Social and Marketing Research Agency.

BRIF (1995b). *Preliminary report on the irrigation rehabilitation project.* Almaty, Kazakstan: BRIF Social and Marketing Research Agency.

Bromley, D.W. (1982). *Improving irrigated agriculture: Institutional reform and the small farmer.* World Bank Working Paper No. 531. Washington, DC: The World Bank.

Bromley, D.W. (1987). Irrigation institutions: The myth of management. In W.R. Jordan (Ed.), *Water and water policy in world food supplies* (pp. 173–176). College Station, TX: Texas A&M University Press.

Euroconsult (1995). *Support for the privatization of state and collective farms and agricultural processing and service enterprises.* Arnhem, Netherlands: Euroconsult, July.

Freeman, D.M. (1991). Designing the organizational interface between users and the agencies. In B.W. Parlin and M.W. Lusk (Eds.), *Farmer participation and irrigation organization* (pp. 35–68). Boulder, CO: Westview.

Freeman, D.M. and Lowdermilk, M.L. (1981). Sociological analysis of irrigation water management: A perspective and approach to assist decision making. In C.S. Russell and N.K. Nicholson (Eds.), *Public choice and rural development* (pp. 153–173). Washington, DC: Resources for the Future.

Fukuyama, F. (1995). *Trust: Social capital and the creation of prosperity.* New York: The Free Press.

Gayle, D.J. and Goodrich, J.N. (1990). Exploring the implications of privatization and deregulation. In D.J. Gayle and J.N. Goodrich (Eds.), *Privatization and deregulation in global perspective* (pp. 5–30). New York: Quorum.

Goetze. D. (1986). Identifying appropriate institutions for efficient use of common pools. *Natural Resources Journal, 27*(1), 187–200.

Karlova, E. (1995). Glossary. In I.W. Lieberman and J. Nellis (Eds.), *Russia: Creating private enterprises and efficient markets* (pp. 245–250). Washington, DC: The World Bank.

Kikeri, S., Nellis, J., and Shirley, M. (1992). *Privatization: The lessons of experience.* Washington, DC: The World Bank.

Korten, D.C. (1980). Community organization and rural development: A learning process approach. *Public Administration Review, 40*(5), 480–511.

Korten, F.F. (1982). *Building national capacity to develop water user associations: Experience from the Philippines.* Staff Working Paper No. 528. Washington, DC: The World Bank.

Lusk, M.W. and Parlin, B.W. (1991). Bureaucratic and farmer participation in irrigation development. In B.W. Parlin and M.W. Lusk (Eds.), *Farmer participation and irrigation development* (pp. 3–33). Boulder, CO: Westview.

Nellis, J. (1991). *Improving the performance of Soviet enterprises.* World Bank Discussion Papers 118. Washington, DC: The World Bank.

Nellis, J. (1995). Introduction. In I.W. Lieberman and J. Nellis (Eds.), *Russia: Creating private enterprises and efficient markets* (pp. 1–4). Washington, DC: The World Bank.

Nellis, J. and Kikeri, S. (1989). Public enterprise reform: Privatization and the World Bank. *World Development, 17*(5), 659–672.

Ott, A.F. and Hartley, K. (Eds.) (1991). *Privatization and economic efficiency: A comparative analysis of developed and developing countries.* Brookfield, VT: Edward Elgar.

Shirley, M. (1989). *The reform of state-owned enterprises: Lessons from World Bank lending.*
    Policy Research Report 4. Washington, DC: The World Bank.
Stoesz, D. and Lusk, M.W. (1995). Social compacts and welfare transformation in Poland.
    Journal of Sociology and Social Welfare, 22(4), 85–98.
World Bank (1993). *Kazakstan: The transition to a market economy.* Washington, DC:
    Author.
World Bank (1995). *Staff appraisal report: Republic of Kazakstan Irrigation and Drainage
    Improvement Project.* Washington, DC: Author, September.
World Bank (1996). *From plan to market: World development report.* New York: Oxford
    University Press.

ENDNOTES

1. For purposes of this chapter, *newly privatized farm* refers to large agro-enterprises
    resulting from the privatization of a state farm, including joint stock companies, collec-
    tive enterprises, comrade associations, and large farm cooperatives.
2. Farm workers are individuals who reside on a newly privatized farm and who participate
    in its collective labor as shareholders or members. They usually function within worker
    brigades which administer subsections of the farm.
3. Farm managers are the directors of newly privatized farms. Elected by the general
    assembly of farm workers or appointed by a board of directors, they are the titular and
    administrative heads of these agro-enterprises.
4. Specialists are technical personnel employed by state farms and newly privatized farms.
    They have training in economics, engineering, irrigation, or veterinary medicine and are
    members and shareholders of the farm as well as employees.

CHAPTER 13

Conclusion: Common Themes and Replicable Strategies

Marie D. Hoff

Communities will not find a standard recipe to follow when they begin to plan for building environmentally sustainable practices in their local area. The wide variety of issues addressed and approaches used by the communities described in this collection demonstrates that opportunities are unique to each setting and that an entry or a beginning point for social change must be found that is specific to the local situation. A quick review of the community cases reveals some essential ingredients, such as the leadership and determination of a core group of people with a vision of how things could function differently, the availability of a reasonable modicum of financial support, and a heightened understanding that plans for economic improvement must be integrated with plans for the development of people and environmental protection, if those economic successes are to be sustainable. Some key factors that helped communities make progress toward sustainable development are summarized in the following paragraphs.

DEMOCRATIZATION OF DECISION MAKING

Democracy is surely one of the golden words of the twentieth century, but even countries with basically democratic governments have had to struggle to develop and maintain effective participatory decision-making structures that involve citizens in collective management of their everyday economic and social life. In Chapter 1, a number of questions were raised regarding how political structures affect the fairness and effectiveness of environmental decision making. Virtually each community described in this volume has wrestled with the problem of building such fair and effective mechanisms for responsible (i.e., sustainable) management of local resources. The transition to democratic structures for husbanding scarce water resources — through development of water user associations in irrigated agriculture — appears to be especially painful to the central Asian country of Kazakstan, as described by Mark Lusk and S.I. Ospanov. The Communist system attempted to obliterate such self-management, and the basics of group decision making are being learned from the ground up. The remarkably successful development on the island of Negros Occidental, Philippines, in David Cox's analysis is traceable to the intensive involvement of extremely impoverished local people in choosing and actually constructing their own local economic resources, such as schools and roads, and reforestation and agricultural improvements.

In the United States, local structures for participatory decision making are also unique to the local situation, but they usually share the common feature of being voluntary organizations, that is, with no formal governance authority. However, the actual influence and effectiveness of such organizations can be considerable if they are broadly inclusive of local, varied voices and if they develop credible, easily understood processes for making judgments and decisions. The Henry's Fork Watershed Council, evaluated by Kirk Johnson, is a widely respected organization in the intermountain western region of the United States. In part, this is because the council's Watershed Integrity Review and Evaluation criteria are clear and publicly known bases for reviewing the merits of development projects proposed by private and governmental organizations. The council's distinctive structure for participation has enabled it to serve as a context for community-level decision making that is genuinely consensual. The typical structure for a voluntary organization in the United States would include membership dues and governance by a policy-making board. However, the council has functioned well without a board, and participation at meetings, led by a facilitation team, is open to all who are willing to observe with goodwill the process for consensus-based deliberations. Michelle Livermore and James Midgley demonstrated that the deep racial divide in the southern U.S. city of Baton Rouge was bridged successfully when the members of the predominantly white Louisiana State University were willing to form a genuine partnership, rather than a paternalistic "helping" relationship, with members of the impoverished African-American neighborhood around it.

A coalition of organizations is another viable model for fostering participation. The Southern Oregon Economic Development Coalition (SOEDC), as reported by Joan Legg and Frank Fromherz, demonstrates two strategies for opening participation opportunities for socially disadvantaged groups that are frequently disenfranchised in the political arena. First, a funding source, such as the Campaign for Human Development, can influence local organizations by requiring the representation of poor and diverse populations in projects which they support. Second, formal requirements in public planning regulations for community representation, such as the SOEDC uncovered and used, serve as a basis to ensure that representatives of disenfranchised groups are nominated and selected for membership on influential public planning bodies.

The sustainable cities movement, a rapidly expanding process in Canada and the United States, exemplifies another important aspect of democratization of decision making, namely, a process for engaging broader and deeper participation by the general population in setting goals and strategies for public policy. Mark Bekkering and John Eyles explained the several public mechanisms for participation which engaged over 1,000 people across the metropolitan community of Hamilton–Wentworth, Ontario, in a process for envisioning how they wanted their community to function by the year 2020. Community-wide participation heightened awareness of environmental issues and strengthened commitment to the public planning goals. In the city of Chattanooga–Hamilton County, Tennessee, as analyzed by Mary Rogge, many people across the metropolitan region also participated in a community-wide planning process. However, poor people's participation in public decision making is developing more slowly and at some considerable cost in the effort they have to put forth to be heard. Social action and persistent demands by African-Americans that toxic materials threatening their neighborhoods have to be remediated have led to cleanup plans which, when accomplished, will bring environmental benefits to the entire community. Confrontational tactics and political action are important routes to social justice and participation where there are significant inequities in political power and economic resources (cf., Fisher, 1994).

RESEARCH AS A BASIS AND STIMULUS
FOR SUSTAINABLE DEVELOPMENT

The majority of communities reported in this volume recognized and emphasized the importance of a solid scientific foundation for environmental restoration and environmentally sustainable economic development initiatives. The Willapa Alliance, as reported by Marie Hoff, adopted good science — and public understanding of that science — as one of its guiding principles for action on local issues. That commitment has supported the organization's credibility, not only with the general public but with scientifically sophisticated decision makers in industry and government. The research process itself can become a channel for social change that is both scientifically sound and socially acceptable when persons affected by the change participate as subjects, rather than objects, of investigation, as portrayed in the research conducted by Jon Matsuoka, Davianna Pomaika`i McGregor, and Luciano Minerbi with the residents of the island of Molokai, Hawaii.

In order to move toward integrating environmental goals with economic development goals, many local communities are using technical research to develop products, or product-development methods, that are environmentally benign and that do not deplete the local resource foundation. In the Willapa Bay region, described by Marie Hoff, research is finding new uses for traditional products, such as cranberries, and how to make better use of previously under-developed and undervalued resources, such as red alderwood. The Appalachian eco-villages, as reported by Jonathan Scherch, have become exciting learning laboratories for a wide variety of sustainable uses of local resources, in the areas of housing, energy, food, and job creation. Likewise, the Yawanawá people, in Sandra De Carlo and José Drummond's report, are developing new commercial uses of plants native to the Brazilian rain forest. Cultivation of these plants occurs on land which was degraded under earlier large-scale deforestation. Other tribal groups in Brazil are eager to learn from their experience.

David Cox explained how decline of the sugar industry on the island of Negros Occidental, Philippines, awakened the villagers to the attendant environmental collapse and aroused their determination to rebuild their agriculture, forestry, and fishing economy on a sustainable, ecological foundation. This kind of experimentation involves the recovery of traditional knowledge and skills and its harmonization with contemporary scientific understanding of good ecological principles for use of renewable resources, such as soil and plants.

Threaded throughout the community studies in this volume is evidence of the importance of social research for sustainable development. First, community-level models of alternative ways of living and working, such as demonstrated by the eco-villages in Appalachia and the revitalized tribal life of the Yawanawá in Acre, Brazil, can also be interpreted as living social experiments. Such natural experiments can be studied, profitably, to glean the elements which can be replicated successfully.

Local research on the community's social history supported sustainable development in Baton Rouge, Louisiana (the Community University Partnership [CUP]) and in Appalachia (intentional community, eco-village development), by uncovering and retaining people's sense of place. The general environmental literature conveys a growing realization that people will care for and defend a place, that is, a local physical environment, when they know, feel, and value their own social rootedness in that unique environment. The challenge to sustainable community development groups is to find creative ways to research and to educate the general public about its local ecology and its social history.

INVESTMENTS IN PEOPLE

As noted in the introduction to this collection, social development practitioners have gained awareness in the past several decades that economic development is more likely to succeed if explicit attention is paid to the human factor. Likewise, if local environments are to be protected and respected for the long-term future, the needs of people also must be considered deliberatively. Investments in human development were a priority in the CUP in Baton Rouge, in the form of literacy education, speaking skills, research skills, leadership and organizational skills, and management skills for small business development. In rural

Kazakstan, skills in self-management, the operational skills of democratic decision making, are essential for movement toward effective and efficient use of scarce water resources and a sustainable agricultural system. The Henry's Fork Watershed Council invested money to train people in consensus-based decision making, rather than simply expecting the skill to evolve spontaneously. The SOEDC invested in training disadvantaged populations in leadership skills, and in economic literacy education, to give people the skills to critique the unjust and unsustainable features of so-called mainstream economic development models (such as tourism and shopping malls). The intentional communities (eco-villages) in Appalachia made a commitment to human skills development by actually developing an Earth Literacy School — a center for teaching the nuts and bolts of alternative, sustainable living, such as knowledge of how to build simple housing, develop wind and solar energy systems, and practice organic agriculture. In Negros Occidental, people also learned skills in sustainable agriculture, forestry, and fishing, but the pivotal skills-training focus that ensured the permanency of the project's development goals was the development workers' emphasis on stimulating the people's internal capacity (i.e., motivation and belief in their own skills and abilities) through participation and control in the activities undertaken. Likewise, Sandra De Carlo and José Drummond (evaluating the level of success of the Aveda Company's project in Acre, Brazil, with the Yawanawá Indians) note that human understanding and educational development constitute key indicators of the likelihood that environmental and economic projects will have long-term viability.

RESTORING AND ENHANCING LOCAL CULTURE

As argued in Chapter 1, unsustainable approaches to the use of environmental resources also tend to weaken the cultural heritage which, over centuries, grew out of distinctive patterns of existing in a particular ecosystem. Culture is a very broad term; here it is taken to encompass virtually all aspects of human social expression, from patterns of family life to artistic expression and community religious and celebratory activities. The participatory research and policy decision process engaged on the island of Molokai, by university-based researchers Jon Matsuoka, Davianna Pomaika`i McGregor, and Luciano Minerbi, enabled the native Hawaiians to not only recover and preserve the traditional subsistence skills of their environmental heritage but also to retain community understanding of how the cultural practices related to use of environmental resources (e.g., for gift-giving and celebrations) are significant in sustaining human relationships and community solidarity. Although the economic success is still uncertain, the venture of the Yawanawá people to learn responsible marketing of products from indigenous plants strengthened and renewed their community solidarity and cultural practices. Their joy in renewal of their community ties to one another is palpable to visitors.

These examples rather ideally portray a key theoretical theme of this collection, namely that human society exists in relationship to, is imbedded in, a particular ecosystem which deeply influences the features of that human society. And reciprocally, the types of social systems influence how a people treat their local ecosystem. The autocratic, centralized decision-making structures under the Soviet political system in Kazakstan, the sugar plantation economic system in the Philippines, and the rubber plantations in Acre, Brazil (where the Yawanawá Indians were virtual slaves) discouraged local ability to care for local resources in a conservative way and destroyed local community relationships in the process. Likewise, adversarial approaches, so typical of policy-making processes in the United States, strangled local stewardship capacity in the Henry's Fork (Idaho and Wyoming) watershed. A new social organization for civic decision making benefited the physical environment. Mary Rogge noted how the unfortunate cultural heritage of racism in the United States contributed to the excessive despoilation of the natural environment in neighborhoods where African-Americans lived in Chattanooga, Tennessee; however, the entire metropolitan area was severely polluted during the heydays of industrial production. Renewal of culture in Chattanooga involved basic quality of life improvements, such as zero-emission buses and improved social services.

The Willapa Alliance, in southwest Washington State, enhances the population's cultural appreciation of its rich local ecosystem and its social history through educational projects, such as illustrated books (Wolf, 1993, *A Tidewater Place*) and computerized programming (*Understanding Willapa*). In the Henry's Fork ecosystem, not only is the local environment better protected, but functional, irenic community relationships are evolving with the development of the Watershed Council, which encourages community conversation, or civic dialogue, about issues of legitimate concern to all. The Watershed Council is an excellent example of how new social structures may emerge as communities struggle with finding better ways to consider the environmental effects of social and economic plans.

The oral history project and restoration of a historic school building in the CUP of South Baton Rouge not only enhanced the personal skills of the participating youth, but assisted the entire community to recover its memory of its cultural heritage. This and other CUP initiatives, summarized by Michelle Livermore and James Midgley, supported the tremendous outpouring of community involvement in neighborhood environmental cleanup of over 64 tons of trash and beautification through tree planting. Likewise, as Joan Legg and Frank Fromherz related the process, when a number of organizations representing or serving disadvantaged groups in Oregon joined together and formed the SOEDC for the good of their community, they considered how to support the community's deep cultural attachment to the local economy's forestry and agriculture, and they also broadened their awareness of and support for the older, rich Native American cultural relationship to the local resources.

An interesting question arises from the study of Narrow Ridge Community and other eco-villages in Appalachia, reviewed by Jonathan Scherch. Such experiments in intentional community are springing up in various localities, primarily in the more industrialized (i.e., wealthier) countries of the world.[1] The question is to what extent such new forms of culture can be created and sustained over time — in comparison to efforts to rejuvenate and sustain extant cultural patterns. Narrow Ridge Community, in rural Appalachia, incorporated an awareness of the importance of historical roots of culture through its efforts to keep alive traditional work skills and tools and its promotion of traditional local celebratory events. It would seem that intentional communities that build on positive features of their larger social context will be more likely to survive and thrive.

The sustainable cities movements, in Hamilton–Wentworth, Ontario, and Chattanooga–Hamilton County, Tennessee, illustrate that people in large, metropolitan areas also cherish the historical story of their community and are working to preserve the positive features of their community's traditions.

SOCIAL NETWORKS

Social networking is a natural pattern of human ways of relating to each other for social and economic purposes (as discussed in the previous section on culture). But, it is also used as a deliberate contemporary social change strategy. Social developers, community organizers, and social workers have researched and utilized social networking for some time (cf., Rubin and Rubin, 1992, p. 170). Recently, social scientists have attempted to capture the notion in the phrase "social capital" (Putnam et al., 1993), namely, that the density and intensity of social relationships among persons in a community are important foundations for economic well-being. An applied approach to building social capital in a depressed neighborhood was accomplished through the resources made possible by the CUP. A wide variety of social networking activities was initiated to stimulate change.

In the communities reported in this volume, social networks were built around bringing people together as individuals (in Kazakstan, Molokai, Brazil, Appalachia); around bringing people together as representatives, or spokespersons, for another specific group or organization (in Baton Rouge, Willapa Bay, southern Oregon); or structuring both of the previous options into their composition and ongoing activities (in Negros Occidental, Hamilton–Wentworth, Chattanooga–Hamilton County, Henry's Fork). Each community deliberately developed or strengthened social networks for attacking local problems and responding to needs. For example, The Willapa Alliance has a board of directors which includes representatives of specific organizations (e.g., Weyerhaeuser Company, The Nature Conservancy) and people who represent a group interest (e.g., oyster growers), expertise (e.g., scientists), or point of view (e.g., a writer/

naturalist). In addition, the Alliance promotes broad citizen participation in its activities through public meetings, organizational membership, and the encouragement of voluntary involvement in community environmental improvement projects. A sophisticated form of social networking occurred in Negros Occidental, in that extensive work went into clearly defining and developing the respective responsibilities of the various local, regional, national, and international organizations that participated in the social development project on the island.

Several authors (Cox, Hoff, Johnson, Livermore and Midgley, Lusk and Ospanov, Rogge) noted the importance of building or restoring people's ability to trust one another as a basis for working together on local issues. In each of these communities, new institutions (i.e., new kinds of social networks) had to be built to create a social space for conversation and communication. It appears that these fresh spaces for encounter helped foster the trust that was essential to find new solutions to old problems.

ROLES OF TRADITIONAL SOCIAL INSTITUTIONS

The need to sometimes build new social organizations and networks does not negate the important role of existing social institutions in building sustainable societies. The communities included in this collection present creative involvement of government entities (Negros Occidental, Molokai, Chattanooga), religious institutions (southern Oregon and Baton Rouge), universities (Baton Rouge, Molokai, Hamilton–Wentworth), business and industry (Brazil, Henry's Fork, Willapa Bay, Chattanooga–Hamilton County), and financial institutions (Willapa Bay, Kazakstan).

Government

As noted in Chapter 1, all levels of government — and their respective roles and responsibilities — must be carefully considered by groups that are crafting sustainability initiatives. As portrayed in Kirk Johnson's study of the Henry's Fork Watershed Council, formally constituted government agencies will not, indeed should not, relinquish their legislated duties to voluntary organizations. Moreover, local ecosystems and local political jurisdictions are imbedded within larger natural and political systems, such that local interests alone may not adequately respond to the common good of the larger whole (McCloskey, 1996). While a blueprint is not proposed, voluntary groups planning a sustainable development project will have to consider carefully how they structure their relationships to various governmental entities. Still another aspect of governmental relationships is how traditionally disempowered groups, such as farmers in Kazakstan, low-income residents in southern Oregon, African-Americans in Chattanooga, indigenous people in Brazil, and native peoples of the islands of Molokai and Negros Occidental, gained a voice in public decision making and fair access to governmentally allocated resources. In other words, the laudable desire for peaceful, consensus-based decision making, under local control, must not obscure or ignore real inequities in power and resource distribution in communities and societies.

Religious Institutions

Religious institutions often play a crucial role in retarding or promoting social change. In most situations, agents of social change are well advised not to ignore religious bodies as an important sector of society. In South Baton Rouge and in southern Oregon, individual religious leaders and religious groups contributed significantly to the material resources and skills needed. More importantly, especially in southern Oregon, an explicit consideration of social values that were shared across religious identity boundaries (Christian, Jewish, Native American) enabled a wide spectrum of community-based, low-income-representing groups to agree to shared goals for environmental and economic development in their region.

Universities

Universities are extremely rich social institutions, in their material holdings (buildings and equipment) and, more importantly, in their reservoir of key knowledge and skills needed by society. Two chapters in this collection report particularly significant contributions of universities to the promotion of sustainable social development. Louisiana State University transformed its relationship of animosity with its impoverished neighborhood to a relationship of equals and thus was able to help stimulate a remarkable range of social, environmental, and economic improvements in the area. Likewise, by using a participatory (i.e., egalitarian) approach to research, three researchers from the University of Hawaii demonstrated that knowledge is indeed a key to power and change.

A number of the communities working to make their way of life more sustainable recognized the importance of solid natural and social science and actively solicited the participation of researchers, typically university based, on their boards and committees. The university plays another important role in assisting communities to evaluate the strengths and limitations of their structures and strategies, as highlighted in Sandra De Carlo and José Drummond's evaluation of the Yawanawá–Aveda Bixa project. Whether through universities or other formal or informal educational institutions, knowledge and education of the general public is an essential foundation for finding sustainable ways to conduct a community's economic business.

Business and Industry

Corporate size and power in many cases exceed that of entire national governments, and thus cannot be ignored if significant progress is to be made toward building not only sustainable communities but a transformed global economic system. Community organization practice typically confronts and challenges inequitable power relations in a community to demand a larger share in power, decision making, and access to resources (Fisher, 1994; Rubin and Rubin, 1992). Residents of Molokai, for example, used this approach with respect to certain demands they made of a large corporation which operated a tourist ranch on the island; residents of southern Oregon successfully challenged the inequitable power of vested business interests in public planning processes in their region. However, as with other strategic tools for social change, confrontation and conflict strategies must be assessed in light of the local situation and used carefully. In Willapa Bay, the environmentalists noted that business interests, notably oyster growers but also cranberry growers, and forestry and fishing interests had strong economic motivation to restore, maintain, and improve the natural environment on which they were totally dependent. Business people played key catalytic roles in developing The Willapa Alliance and a related environmentally oriented financial institution, the ShoreTrust Trading Group. In the Henry's Fork Watershed Council, residents with business or environmental protection interests found that consensus approaches forwarded the goals and agendas of both groups better than had the previous adversarial conflict model for influencing policy. In Chattanooga–Hamilton County, the early action by industries and the Chamber of Commerce was a crucial factor in the area's rapid progress, particularly in simultaneously promoting environmental and economic interests.

Although the Aveda Company's project with the Yawanawá Indian tribe for the production of cosmetic dye had mixed success, it typifies an important new thrust, the "greening" of industry, that is, the efforts by many businesses to become more environmentally responsible in the way they do business (cf., Hawken, 1993; Stead and Stead, 1992). Although some businesses may engage in "greenwash" (a public relations veneer), there is a hopeful sustainability ethic developing among many businesses.

CONCLUSION

This set of community studies does not purport to be either an ideal or a comprehensive survey of sustainable development projects occurring around the globe. The cases reported here also do not represent finished projects. Each community faces enormous challenges for the future. Each community's struggle toward inventing and discovering social and economic practices which honor the natural limits of the ecosystem is a work in progress. But there are many excellent lessons to be learned from the wide variety of sustainable community development experiments reported. More importantly, I hope these communities' varied stories of quest for a sustainable way of life have inspired and motivated you, the reader, to take action in your own local region and community.

REFERENCES

Fisher, R. (1994). *Let the people decide: Neighborhood organizing in America.* New York: Twayne.

Hawken, P. (1993). *The ecology of commerce: A declaration of sustainability.* New York: Harper Business.

McCloskey, M. (1996). The skeptic: Collaboration has its limits. *High Country News, 28*(9), 7.

Putnam, R.D. with Leonardi, R. and Nanetti, R.Y. (1993). *Making democracy work: Civic traditions in modern Italy.* Princeton, NJ: Princeton University Press.

Rubin, H.J. and Rubin, I.S. (1992). *Community organizing and development* (2nd ed.). New York: Macmillan.

Stead, W.E. and Stead, J.G. (1992). *Management for a small planet: Strategic decision making and the environment.* Newbury Park, CA: Sage.

ENDNOTE

1. See the journal *YES! A Journal of Positive Futures* (a new magazine which is the successor to *In Context*) for an ongoing chronicle of experiments in intentional community building, as well as regular reporting on other aspects of building new cultural practices for a humane and environmentally benign society (available from Positive Futures Network, P.O. Box 10818, Bainbridge Island, WA 98110; e-mail: yes@futurenet.org). See also the intentional community network available at www.ic.org.

Index